Non-linear Theory of Elasticity and Optimal Design

Non-linear Theory of Elasticity and Optimal Design

How to build safe economical machines and structures
How to build proven reliable physical theory

Leah W. Ratner

2003

ELSEVIER

Amsterdam – Boston – Heidelberg – London – New – York – Oxford
Paris – San Diego – San Francisco – Singapore – Sydney – Tokyo

ELSEVIER B.V.
Sara Burgerhartstraat 25
P.O. Box 211, 1000 AE Amsterdam, The Netherlands

First edition 2003
ISBN: 0-444-51427-9

⊗ The paper used in this publication meets the requirements of ANSI/NISO Z39.48-1992 (Permanence of Paper).
Printed in Hungary.

Preface

Contemporary physics from its beginning in the 17th century has been progressing in two parallel directions, i.e. empirical and mathematical. One unresolved problem of epistemology is that these two branches are not really well combined in the scientific theories. Each of them has its role. The empirical methods are used for governing the facts concerning a phenomenon and testing the inferences of a hypothetical theory. Mathematical methods are used for description of the hypothetical physical ideas and making mathematical inferences from these hypotheses. The assumption is made that the mathematical inferences can be tested empirically and that such tests may perhaps not prove but at least validate the theory. Nevertheless empirical validation is insufficient for combining the methods. There is no proof for theories that remain in essence hypothetical. The successful alignment of essentially different methods can be achieved by employing logical structure as a mediating method.

In this work the author proposes logical structure as the frame of a physical theory that allows building a consistent provable theory. The theory presented in this work is a Non-Linear Theory of Elasticity. This theory has a logical frame that makes it a reliable foundation for structural analysis and design.

Part I of the book describes the general principles on which the non-Linear Theory of Elasticity was built. The theory has a new conception of strength and elastic stability of a structure. This part also reproduces the specification of the author's US patent on the method of optimization of structures.

Parts II and III are devoted to the analysis of the current Linear Theory of Elasticity and the new Non-Linear Theory of Elasticity. Part III also analyzes typical structures such as bars, beams, shafts, columns, plates, and shells. The reader will also find there the important discussion on the distribution of elastic forces in a structure and a new hypothesis on the torsion of non-round bars.

Part IV considers some important methodological questions relating to the construction of a theory, such as graph analysis and the geometrical models of physical functions.

Part V discusses philosophical implications of the new methodology in science and discusses in length the definitive logic in the theory of elasticity. The important physical implication of this methodology is the need for a mathematical description of the domain of stable physical relations for a physical phenomenon.

Contents

Introduction

> It is one of the boasts of modern science that it is a truly open-ended intellectual system in which dissent is both welcomed and rewarded. The practitioner has been brought up on this idea and proudly repeats it until, perhaps, he finds himself on the side of dissent. Then the "open" ranks suddenly close and he finds himself isolated and alone, wondering how it happened that his careful adherence to the rules of the game has led to ostracism. *(L. Pearce Williams)*

"Non-Linear Theory of Elasticity and Optimal Design" deals with developing and proving a new fundamental theory. Although the useful concepts and methods of the current Linear Theory of Infinitesimal Deformations remain, the basic physical concept of strength and elastic stability of a structure changes. The logical structure of the theory of elasticity, the concepts, the criterion of strength and elastic stability, the equation of deformation have been changed, and an equation for elastic stability was added. The method for optimizing the dimensions of a structure is new. The approach to mathematics in physical theory is changed. A new point of view on the role of logic in the construction of physical theory is presented. Logic becomes definitive.

The theory of elasticity is the foundation of structural design. An important characteristic of elastic relations is the limit of elasticity. The limit for an individual structure currently can be found only by testing the structure destructively. The reason is that linear theory by its nature cannot describe a limit, because a limit is not a property of a linear function. The non-linear theory presented in this book, on the other hand, describes limits for individual structures and allows optimization of structures.

A new concept of strength is associated with the non-linear theory of elasticity. The actual limit of elasticity of a structure, which reveals itself in the destruction of the structure, can be of different physical origins. It can be the limit of the material, but more often it is generated by the geometry of the structure. Both limits should be known for structural analysis and design to be successful. This book describes a simple non-destructive method of establishing minimal reliable dimensions of a structure.

This engineering problem is at the foundation of structural analysis and design. The safety and cost of a structure in the mechanical, civil and aerospace engineering fields depend on establishing minimal reliable dimensions for the structure. The problem, formulated in 1638 by Galileo, "is to find the form of the generating curves so that the resistance of a section may be exactly equal to the tendency to rupture at that place." Galileo was unaware of the elastic properties of materials and did not describe relations mathematically.

The English physicist Robert Hooke discovered the existence of elastic properties of materials and structures in 1678. Since that time a continuous effort has been made to find a scientific method for predicting the limit of elastic relations and establishing safe dimensions of structures within that limit.

There are reasons for the fact that this optimization problem has not been solved although a mathematical method of optimization exists. If the problem could be solved using empirical, statistical and probabilistic methods, it would have been solved already, for there is no lack of empirical data. If it were possible to find a solution for the optimization problem within the framework of the established linear theory of elasticity, then it would have been done in the 19th century when the mathematical apparatus of linear theory was developed.

Solving the problem of structure optimization is possible only after a revision of the linear theory of elasticity, its logical structure, mathematical apparatus and physical foundation, which are presented in this book. Solving the problem is connected not only with the criticism of current theory, but also with the development of a new reliable method of construction and verification: the Non-Linear Theory of Elasticity (NLTE).

The new theory has a new criterion that designates fundamental changes. "If the actually formulated laws of our physics can be shown to undergo change themselves, it can only be in reference to something else which is constant in relation to them" ("A Preface to Logic", Morris R. Cohen). Currently, the criterion for design calculations is the limit of elasticity of the material. NLTE factors in the rate of change of deformation. The main reason for choosing this criterion is that the limit of elasticity corresponds with a rapid increase of change in deformation. A mathematical description for the rate of change is missing in the linear theory. The new equation of elastic stability is obtained as a derivative of the basic equation of deformation. Not every description of deformation reflects elastic relations correctly. Here a new equation of deformation is presented and justified.

In the 20th century new technology to build high-rise buildings, airplanes, bridges, and the like, developed rapidly. But at the same time the science of structural design stagnated. Engineering disciplines such as "Strength of Materials" and the "Theory of Elasticity" have practically been closed to free independent scientific thought by standard-setting organizations that control scientific ideas and research, engineering publications and engineering practices. For example, one of the prominent standard-setting organizations, the American Institute of Steel Constructions (AISC), representing the interests of steel fabricators, has served as the link between the steel monopolies and the countless manufacturers and builders who use standard steel products. Scientific laws, which have a tendency to change, became the objects of governmental laws. Another reason for the slow development of science is the inertia of established theoretical principles.

Until now the theory of elasticity has been designed as the Linear Theory of Infinitesimal Deformations. The crisis of linear theory came to light at the end of the 19th century after the main principles and mathematical descriptions of the theory had been developed. "Except in very simple cases, the demonstrations are less rigorous than those which form the Mathematical Theory of Elasticity, an exact science which is unable to furnish solutions for the majority of the practical problems which present themselves to the engineer in the design of machines and structures." ("Strength of Materials", Arthur Morley, Eds. 1908 till 1954). Since the 1950s no major changes in this science have occurred despite the fact that the design process has become more complicated, uncertain and expensive.

Under pressure of the demands of steel construction technology and new ideas that have infiltrated the field, the AISC recently changed its manual and specifications. An allowable stress design (ASD) specification was substituted for the load resistance factor design (LRFD) specification. However, according to the AISC, in the new specification the "philosophy of design remains the same." In the "Steel Design Handbook", edited by Akbar R. Tamboll © 1997, the reason given for a new method is that "until recently engineers were basing the analysis and design of structures on a linear theory of elasticity. On the whole, the results have been satisfactory. The buildings and bridges have withstood the test of time. Why then should one be concerned with the LRFD method? Finally, elastic analysis of all but the simplest of structures is complicated. Obviously the net result is a waste of material. For structures such as aircraft, where weight is of prime importance, the results may be even more serious. Further, since such an analysis would have little rational basis, a true estimation of the safety factor would become virtually impossible." This explanation still contains no rational basis for the LRFD method.

No theoretical foundation has been offered for the LRFD method. It has been maintained that the LRFD specification accounts for the factors that influence strength and loads by using a probabilistic basis and statistical methods. However, the probabilistic basis and a rational logical–mathematical deterministic basis are two different approaches. A statistical method has a rationale for its use when a reliable mathematical description of the elastic relations exists and statistics give the deviation of the empirical data from the mathematical description. Without sound theory a statistical method is just speculation. The problem with the linear theory of elasticity is that it distorts the relations it describes. No amount of statistical data and probabilistic method corrects that.

The need to revise the linear theory arises from the fact that this theory is fundamentally inconsistent with the experimental observations. Thus, linear theory identifies the elastic limit for a structure with the limit of elasticity of the material, while observations and experiments show that different structures made of the same material have significantly different limits depending on the geometry of the structure. In physics such disagreement raises doubt in the theory.

The Linear Theory of Infinitesimal Deformations is based on the assumption that because deformations are very small in comparison with the dimensions of a structure, the relationship between them can be described with a linear function. Here we will consider the Non-Linear Theory of Elasticity. One of the physical points of view in the new theory is that although deformations are small we should nevertheless make the necessary comparisons among them in order to detect the rapid changes that describe the limit of elasticity. Linear theory by its nature cannot describe changes in the rate of deformation, because the rate is a constant in a linear function.

The rules of logic are well suited for the task of construction and proving NLTE. This book contains extensive study and analysis of current approaches in the logic of science, as well as a new approach that proves to be constructive for NLTE and can be useful for the other branches of physics and for science in general.

For a successful engineering practice we need a consistent self-proving theory. Modification of the linear theory with formulas that have no foundation in the theory does not improve the theory or the design process. New principles that are logically, mathematically and physically justified have been known for a number of years. The National Science Foundation, the Energy Department (National Bureau of Standards), the AISC, the American Society of Civil and Mechanical Engineers, Argonne National Laboratory, and numerous scientists in the field of structural design have evaluated the new theory. The new theory and the method of optimization have encountered not one scientific objection. The method of optimization of structures has been issued a US patent. And, most importantly, the method includes a non-destructive experimental part to prove itself each time one uses it. There is no particular method of design that can survive without using the new knowledge of non-linear theory of elasticity. Overall the new theory and method can result in significant savings of materials, energy and engineering time. The method of optimization is a powerful analytical and experimental tool for structural analysis and design.

Prologue

The author has two main purposes: to develop a non-linear theory of elasticity and to advance the logical apparatus for the new theory.

Initially the author's main concern was the development of a method that would allow engineers to do comprehensive and reliable design of safe structures with the optimal correlation of dimensions. Then it became apparent that generalization and proof of such a method requires profound changes in the theory of elasticity. These changes in turn show that the logical structure of non-linear theory of elasticity departs fundamentally from the current principles of logic. The current principles are insufficient for constructing and proving a physical theory. The principles of logic in the non-linear theory of elasticity are explained and generalized for the first time in this work.

In my engineering practice I solved some specific problems of optimization of structures, at the same time realizing that the linear theory of elasticity does not provide means for optimizing structures. For instance, for a series of conical polishing tools with slit I had to find a correlation between cross-sectional and longitudinal dimensions for achieving the best springing properties for a series of tool sizes. I solved the problem by comparing the property that later was defined as geometrical stiffness of the particular tool size in the series with the corresponding property of a tool that was found to work satisfactorily. Although I solved this particular problem, the next optimization problem, for a different structure, again demanded creating a method and a criterion for comparing the elastic properties of similar structures. The conclusion was that until the engineer has a suitable optimization theory the solution of each individual problem will be a creative process that is lengthy and not necessarily successful.

In my search for a method of optimization it became clear that first of all we need a criterion for comparing the elastic properties of structures rather than some fixed criterion for the elastic properties of a material. The new common criterion for calculation was assigned to the rate of change of deformation. The rate of change of deformation is associated with the elastic behavior and elastic failure of a structure. The method of optimization was invented by establishing a criterion for comparing structures that is defined as the coefficient of elastic stability, by defining a characteristic called geometrical stiffness that describes elastic geometrical properties, and by describing the elastic relations that would reflect relations for a set of similar structures.

The problem with the construction of a new equation of deformation was psychologically connected to the false belief that if an equation describing physical relations is supported by numerous empirical facts it is, probably, a proper description. The realization that logical and physical correctness is different from mathematical and empirical correctness was the next big step on the road to the new theory. As a result of the more critical attitude to the mathematical procedures in physics a new equation of deformation was constructed. The equation becomes part of a logical structure that can be proven.

The limit of elasticity is associated with a rapid increase in deformation. The identification of the limit of elasticity of a structure requires a derivative equation describing the rate of change of deformation. This change of criterion by itself means that we are dealing with a new theory. If a derivative equation is to be explored the basic equation has to be non-linear. In contrast to the current linear theory of elasticity the new theory was named Non-linear Theory of Elasticity.

The experiment that was designed to test the new equation of deformation not only demonstrated that the method of comparative analysis of similar structures allows one to calculate and anticipate the individual limit of a structure, but also showed that there exist two limits of different origin. The real limit of elasticity can be either the limit of the material or the limit generated by the geometry of the structure, depending on which is smaller. The conclusion followed that the real limit of a structure is relative in nature. This is a new physical foundation of the theory of elasticity.

The way the theory was constructed also appears to be new in the methodology of science, and requires explanation. Analysis of the new theory showed that the logical principles of selecting the data, selecting a proposition, making an inference, establishing the domain and proving the logical structure are new as well. For selected initial conditions this logical structure allows us to make an inference that can be proven correct.

For the construction of a mathematically proven physical theory we have to find a known mathematical analog for the physical behavior in question. The deductive and inductive methods of logic were redefined with the purpose to attain certainty of logical judgment. Certainty of deductive inference relies on strict rules of deduction. Certainty of inductive inference is found under the umbrella of accepted universal laws that permit making logical inferences. Both methods are employed in constructing the statements and logical structure of the non-linear theory of elasticity. Those statements require a validation. For this purpose we use empirical and mathematical methods. Mathematical verification applies to the calculation of unobservable terms, calculations within a function, and the building inference of the described relations. Empirical validation applies to the observable terms that can be measured and to testing the results of inference in the interval of obvious changes.

Better differentiation of the areas of application of deductive and inductive methods of logic and the proper use of empirical and mathematical methods of validation allow one to build and prove physical theory.

Part I. Principles and Methods of NLTE

1. Practical problems

Design of a structure in the mechanical, civil and aerospace engineering fields is associated with specific engineering problems. Here are some of them:

(1) It is a time- and money-consuming process to establish safe and reliable dimensions to withstand structural forces. Currently, even with extensive and expensive technical calculations and destructive tests we cannot determine whether a particular structure can be made lighter, stronger and more predictable in a working environment. The new method can provide such analysis and we can inexpensively achieve optimization of the dimensions.

(2) Often manufacturing products come in a series of similar structures distinguished by their dimensions. Currently there is no method for comparing such structures. Even the idea of necessity of comparison of similar structures for the selection a structure is not part of current practice and had not appeared in the literature until 1986 when an article by the author appeared in "Machine Design". An expensive research and development project for designing one structure does not guarantee the success of similar products. The new method allows for such comparison. The result obtained in a non-destructive test on one structure can then be used for optimizing a series of similar structures.

(3) Currently, no non-destructive method can predict limiting stress and deformation for a real structure. The new method allows for calculation or use of a non-destructive test to predict the individual limit of a structure.

(4) The new method allows for calculation of a mean value of structural forces when the external forces are unknown.

(5) The new method allows for calculation of the optimal correlation between length and cross-section. Note that although the proponents of LRFD (1994 specification of AISC) said that it is possible to establish such correlation with LRFD, they failed to explain how.
(6) Currently there is no common theory for the construction of different structures. Many different theories exist, such as for beam design (statically determinate and statically indeterminate), column design (short column and long column), shaft design (round cross-section and non-round cross-section), plate design (thick plates and thin plates), shell design, cylinder design.
(7) Linear theory has no reliable and comprehensive method for structure optimization. The majority of structures are overweight, some structures are underweight. Both problems, especially in combination, present serious obstacles to safety of design. Overweight parts of machines and structures demand similarly overweight supports. Vibration energy depends on the vibrating mass; overweight parts are better transmitters of vibration to other parts of a structure or to adjacent structures, which might not be designed for this propagation of vibration energy.
(8) Other negative effects of the tendency to construct overweight structures are economical and environmental. Comparison of the results of calculating dimensions with the linear and non-linear theories lead to the educated judgment that structures calculated with the standard method are at least 30% overweight. This means that for the same structure we need 30% more ore, more processing, more transportation, more metalworking machines, and more energy, resulting in more pollution of the environment.

2. Foundations of the non-linear theory of elasticity

Analysis of the sources of inadequacy of the linear theory of elasticity has led to the justification of the Non-Linear Theory of Elasticity. From a practical point of view, linear theory provides the equations needed to calculate the deformations and stresses in a structure. The theory also provides the mathematical description of the geometry of the structure.

But the linear theory has no mathematical means for analyzing the values of deformations and stresses or for analytically determining limiting deformations and stress. The theory has no criterion for comparing the elastic behavior of similar and different structures, and thus provides no means for the generalization of experimental data.

The principles underlying the non-linear theory arise from a body of existing scientific knowledge. Inferences from this knowledge serve as the foundation for the new knowledge and the new principles. The Non-Linear Theory of Elasticity is constructed in accordance with certain logical principles that facilitate the solution of practical problems. The new principles are presented in axiomatic form; these principles are comprehensive. Let us consider the main principles and their practical consequences.

(1) If physical relations have a limit, then these relations require a description with a non-linear function. This demand has several physical and logical reasons. Elastic relations have a limit. Although the deformation, elastic force, geometry and material relations have the limit of elasticity of the material as a boundary condition, these relations also have an observable limit that depends on the geometry of the structure. This structure-specific limit usually is the actual limit for a structure. The continuous function describing deformation, elastic force, geometry and material relations includes the limiting elastic deformation, be it the limit of a material or a specific limit generated by the geometry of a structure. It is necessary to have a mathematical method to detect the limiting deformation.

The limit of elasticity is characterized by an increase in the rate of change of deformation. The linear theory of elasticity operates with a linear function of deformation. Such a function does not differentiate the change of deformation adequately for identifying the limit: the derivative of a linear function describing a rate of change is a constant. A standard function of deformation is shown in Illustration 2 (below). Below the limit of elasticity of the material, there is no clear description for the individual limit of elasticity of a structure. It can be at any point of the line within the possible range of elastic relations. Until now a destructive test has been employed to find the real limit of a structure. If the elastic limit is to be found, and is to be determined mathematically or with a non-destructive test, the description of the elastic relations should be made with a non-linear function, such as that shown in Illustration 3. Differential calculus then enables us to find the limit of elasticity in the interval of rapid changes of deformation by using a derivative equation that describes the rate of change of a function. The practical demand of finding the limit of elasticity of a structure leads us to the conclusion that we need a non-linear description of elastic deformation. The mathematical means should be adequate for serving the physical purpose.

(2) The physical nature of the theory of elasticity and its practical purposes require verification of its mathematical descriptions. Verification, according to current norms of logic, refers not so much to verification of a proposition as rather to the confirmation that the inference from the initial hypothetical description is true to the facts.

A mathematical inference from a basic function is the derivative function. In order to conduct such a test, we should have a logical–mathematical system consisting of a basic equation and a derivative equation. One of the disadvantages of the linear theory of elasticity is that a linear basic function cannot produce a meaningful derivative function: the derivative of a linear function is a constant that does provide knowledge about changes of deformation. We have no way to test linear theory. On the other hand, a non-linear theory may have a derivative function and, in principle, can be verified or proven false. The logical demand of verification of a theory also requires a non-linear description.

(3) Then, why choose a linear function to describe elastic relations? There is no logical reason for such selection. Mathematical descriptions in physical theory are linked, as a rule, to our hypothetical physical ideas about phenomena rather than to logic.

The physical idea behind the linear theory of elasticity is that deformations are so small in comparison with the dimensions of a structure that the relation between them can be represented by a linear function. The theory was even named the Linear Theory of Infinitesimal Deformations. However, the concepts "small" and "large" are relative. Elastic failure occurs within a small range of deformation. The limit of elasticity is associated with a rapid increase in the rate of change of deformation rather than with the absolute value of deformation.

In order to see and appreciate changes of deformation an analysis must be done at the level of the values of deformation rather than at the level of the dimensions of a structure. Mathematically such analysis is possible if the selected independent variable relates to deformation describing the *effect* of the geometry of a structure on deformation. The independent variable concept has to be a function that allows us to make choices. This variable concept may have values comparable with the values of deformation. Then the function of deformation becomes non-linear and may offer an opportunity to evaluate changes of the elastic deformations, describe the limit and test the description.

(4) Elastic relations are physically definitive and so should be the description. The definitive description includes basic and derivative functions. The basic equation gives the absolute values of a function corresponding to certain values of an independent variable concept. The derivative equation gives the relative position of each point in the continuous function and thus describes the elastic behavior depending on this position.

The need for considering the derivative function for the definitive description of structural behavior is clear in the illustration of two functions as curves in Illustration 1. Each point on a curve in a plane is characterized by the absolute values of its coordinates (x, y) and by its relative position on the curve, $dy/dx = -\tan \alpha$. If point A (x, y), which is a structure with the geometrical characteristic x or $f(x)$ and the corresponding deformation y, belongs to function #1, then the structure is in a position of elastic stability with very slow possible changes of deformation. If however point A (x, y) belongs to function #2, then it is in a position of elastic failure where small changes of the variables produce a rapid increase of deformation. In the real physical world only one behavior is true for the given structure and thus only one description is correct.

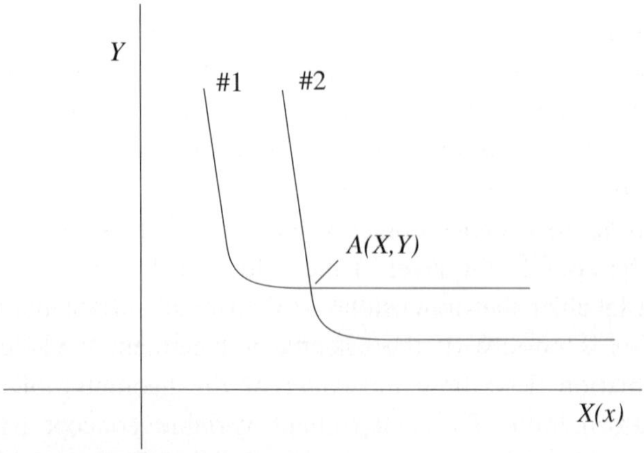

Illustration 1. The behavior of a structure A (x, y) is determined only if dy/dx is known, for the behavior depends on the position "A" occupies on the curve. Structure "A" is in the position of elastic stability if "A" belongs to curve #1. Structure "A" is in the interval of elastic instability if "A" belongs to curve #2. Knowledge of the absolute values (x, y) is insufficient for evaluation of a structure. Definitive description should include a basic function and a derivative function.

In order to find which of the possible descriptions is correct, it is necessary to examine the equations at both levels, for the distinction is revealed in the derivative equation.

(5) From a number of descriptions with the same physical variable concept, such as $Y = f(X)$ and $Y = f(F(X))$, only one description can be physically correct. In mathematics, a function of a function can be presented as a single function. In physics, such transformation can be a cause of a fundamental misconception. The functions obtained in such a manner are not identical physically. Though in both functions the same value of x produces the same value of Y, these equations have different derivatives depending on whether x is an independent variable or it is a variable part of another independent variable concept, for this selection implies different physical behavior. For example, deformation described with the standard equation of elongation $e = NL/EA$ for the independent variable A (area of cross-section) gives the derivative function $de/dA = -NL/EA^2$. The value $\tan 45° = -1$ corresponds to $NL/EA^2 = 1$, and $A = \sqrt{NL/E}$. Let us consider the description

$e = N/ER$ in which $R = A/L$ with the constant length L. The independent variable R represents the characteristic of geometrical stiffness. The derivative of the new equation is $\mathrm{d}e/\mathrm{d}R = -N/ER^2$. The value of deformation in both equations is the same. However, the cross-sectional area of a structure corresponding to $\tan 45° = -1$ indicating a certain change of rate of deformation in this case is different, $N/ER^2 = 1$, $R = \sqrt{N/E}$, $A = L\sqrt{N/E}$.

Physical determinism requires selection of one correct description with the exclusion of other descriptions. In order to make such a selection we have to examine the two-level mathematical–logical structure – the basic function and the derivative function.

Mathematical transformations in physics have some restrictions, because unlike a mathematical function, a physical function describes concrete relations among the variable concepts that have concrete physical contents and meanings. The two-level logical–mathematical system of equations in physical theory needs verification.

(6) In classical logic, verification is a process of confirmation the inference from the initial hypothetical description by the corresponding facts. However, theoretically and practically the confirmation of a physical theory cannot be achieved with the empirical verification of an inference. We deal with two types of equations. The basic equation operates with the absolute values of the variables, $Y = f(X)$. The derivative equation, on the other hand, has the relative value as its result, $\mathrm{d}Y/\mathrm{d}X = f'(X)$. The absolute values can be measured and the basic equation can be tested experimentally. Thus, in the equations $e = NL/EA$ and $e = N/ER$, each of the variables can be measured. The resulting deformation e can be measured and compared with the calculated value. However, the results of the derivative equations $\mathrm{d}e/\mathrm{d}A$ and $\mathrm{d}e/\mathrm{d}R$ are the mathematical relative values. They can be calculated but not measured. The derivative is correct if we described the relations in the basic equation correctly. We are in a circle of a logical indeterminate system.

There is no means to prove the logical structure "If proposition P is correct, then conclusion Q is correct" experimentally. In the theory of elasticity, experimental support for the derivative can be found in the interval of rapid changes by calculating the derivative corresponding to elastic failure where the physical changes are obvious. Even then the nature of a limit needs evaluation, i.e., does it really belongs to the description or is it the limit of the material? Still, we have no means of proving the derivative values for the other points of the physical function. Such limited support is not sufficient for proving that the function is correct.

In principle, it is impossible to measure the relations, and a derivative equation describing a relation is not in its whole experimentally testable. Then, for testing a theory we should examine the system of the equations rather than testing inference.

(7) The objective mathematical tool for testing a hypothetical function can be the derivative function of the universal known properties, for example, the tangent function. The tangent function describes changes that are also characteristic of the changes of deformation, i.e., small changes in the interval $0 < Y < 1$ and rapid changes in the interval $1 < Y < 10$. The system consisting of the basic function and the derivative function is correct when the derivative function is a universally known function that does not need verification, and the basic function that corresponds to that derivative can be verified experimentally. In particular, if the continuous non-linear function of deformation occupies the domain of a tangent function and its derivative at any point is the constituent part of a tangent function, then such physical function is uniquely determined.

Note that we are not speaking about a derivative describing a change of a hypothetical function, but about constructing a function corresponding at any point to an independently known derivative and then testing such function experimentally.

(8) Let us consider how this logical demand changes the theory of elasticity. Numerical analysis of the standard equation of deformation $e = NL/EA$ shows that this function within the limit of elasticity of material exists only in the interval of slow changes (Illustration 2). The function is reduced to a linear function because the values of deformation are several orders of magnitude less than

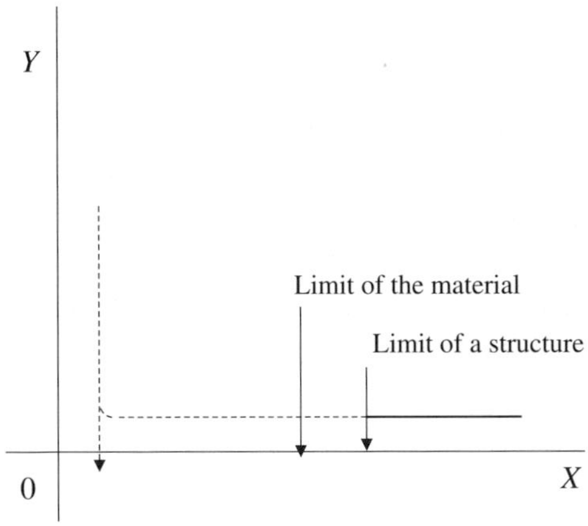

Illustration 2. The standard equation of deformation, for example $e = NL/EA$, within the limit of material is the linear part of the function, which has no mathematically defined limit. Such function does not differentiate changes of deformation sufficiently and does not describe physical relations adequately.

the values of the selected independent variable that is the area of cross-section.

Linear presentation of the relations implies that no limit follows from the relations described in the equation. But this mathematical conclusion contradicts the physical findings, i.e. that the limit in the majority of cases originates in the geometry. The mathematical description of the relations with a linear function is incomplete. It cannot be completed and tested.

On the other hand, selecting geometrical stiffness $R = A/L$ as an independent complex variable concept allows for verification of the function $e = N/ER$. Its derivative $\mathrm{d}e/\mathrm{d}R = -N/ER^2$ may form a tangent function. The tangent function has two intervals: the interval of slow changes, $0 < \tan\alpha < 1$, and the interval of rapid changes, $1 < \tan\alpha < \infty$. The new function may exist in both intervals of tangent function. The variable R can be selected in the interval of rapid changes. It gives the mathematical possibility to calculate the limit of elasticity and to test it experimentally.

The continuous function that exists in the interval of rapid changes necessarily exists in the interval of slow changes. Thus, the new function of deformation is completely defined mathematically by its derivative. The rate of change of deformation can be a criterion for calculating geometrical stiffness of a structure with the desirable behavior.

The experimental confirmation of the system of basic and derivative functions can be found by testing what has been constructed. The new equation of deformation in this case gives the same resulting deformations as the standard equation of elongation, whose results were validated experimentally. Also, the limit of elasticity that is predicted with the derivative equation may have experimental support in the interval of rapid changes of deformation.

(9) The patented method of optimization of structures was developed based on this possibility to predict the limit of elasticity and to predict the elastic behavior of structures with a necessary degree of mathematical certainty. The new equation of deformation, the differential equation and coefficient of elastic stability are the mathematical means of optimization.

(10) The description of elastic relations in both intervals of tangent function has led to the conclusion that there is an individual limit of a structure that depends on the variables in the function of deformation. The experiments support this conclusion. The experiments also showed that the mathematically obtained limit is correct only within the limit of elasticity of the material. The limit of elasticity of a structure that reveals itself in the destructive test is of a relative nature. It can be the limit of the material, or it can be the limit generated by the geometry of a structure. The actual limit of a structure is of comparatively lesser value. For practical purposes we need to know both limits.

2.1 Summary

The linear theory of elasticity is an inadequate description of the phenomenon, for it cannot provide a description for the limit of elasticity and cannot predict the elastic behavior of a structure. Linear theory also has no mathematical means to prove its validity. In order to prove or disprove a hypothetical function it is necessary to test the inference from this function. A linear function has no derivative function that completes description of the physical relations. The Non-Linear Theory of Elasticity, on the other hand, can provide a complete adequate description. The complete description has a mathematical–logical system consisting of a basic equation and a derivative equation.

However, this system cannot be proved by means of experiments only. The derivative equation has a mathematical relative value as its result and therefore cannot be verified experimentally. In order to build a reliable system of equations, the derivative should be a universal function such as a tangent function. The variables need to be selected to satisfy the equations at both levels. The basic equation that corresponds to such a derivative can be, and needs to be, proven experimentally. Such a mathematical–logical system is consistent, proves itself, and excludes other descriptions. The Non-Linear Theory of Elasticity allows us to solve a majority of practical problems with a degree of mathematical certainty. It is a general theory for different structures.

2.2 Recapture

(1) Elastic relations have a limit that depends on the geometry of a structure. The limit characterizes the increase in the rate of change of deformation. A non-linear function describes changes in the rate. A linear function has a constant rate. This property of the functions suggests the selection of a description of elastic relations with a non-linear function.

(2) The physical function ought to be proved. For testing a function we need mathematical inference. Only a non-linear function can give a meaningful inference.

(3) Verification of a description of the elastic relations can be achieved only if the basic function occupies the domain of a tangent function in the interval of rapid and slow changes. Then, the limit of elasticity can be found in the interval of rapid changes, and the behavior of any structure is defined in the full range of function of deformation.

(4) For the construction of such a verifiable system of non-linear functions we selected an independent variable also as a function that can exist in the interval of significant figures, $1 < dY/dX < 10$. This interval is characteristic of the rapid changes.
(5) The logical structure as a basis of physical theory, which is presented here, allows for the construction of a mathematical description that gives objective physical results.
(6) The new non-linear equation of deformation, the derivative equation of elastic stability and coefficient of elastic stability $C_s = [\tan \alpha]$ allow for optimization of structures.

3. Devising the non-linear theory of elasticity

The purpose of this work is to develop a theory of elasticity that makes it possible to predict the elastic behavior and failure of structures. In order to make reliable predictions we must establish a logical connection between theory and physical phenomenon.

Let us consider the logic tools that we have for the construction of a physical theory. First, there is the basic logical argument, "If proposition P is correct, then necessarily conclusion Q is correct." Second, "A universal proposition (P) is verified when its particular consequences (Q) are found to hold true in experience." Thus, one cannot prove hypothetical premises without proving necessary inferences. But inferences are correct only if premises are correct. As a remedy for such circular argument, formal logic offers an empirical rather than logical path to knowledge. "From certain initial data, derive a hypothesis. Deduce from its assumed truth one or more implied facts not included in the initial data. These are its "implications". Try by luck or cunning, observation or experiment, to obtain more findings, which agree with the deductions, made from hypothesis. Such subsequent data confirm the hypothesis" (from "College Logic", Alburey Castell). The current theory of elasticity does not address even these vague suggestions; there are no mathematical implications and tests in the linear theory of elasticity.

One way to approach the problem is to think of our theories as superstructures, which are based on the discoveries of the laws of nature.

Such an approach is based on faith in our ability to understand the connections in nature and to seek empirical support for them. The other way is to build a theory that has a physical basis and a logical–mathematical structure as two equal but independent parts of the theory. The logical structure is included purposefully as the framework of the theory. Physical concepts are selected to satisfy the logical structure. This approach allows us to establish the logical connection between a theory and the phenomenon it describes.

Let us consider the necessary logical structure of a physical theory. In physics, a law is given in the form of a function. The function can be presented analytically as an equation and graphically as a curve. Considering the properties of a curve we can see that each point of the curve on a plane is defined by the absolute values of the coordinates (x, y) and with its relative position on the curve. The relative position is characterized by the tangent at this point, which gives the rate of change of the function. This is a very important characteristic for the prediction of physical behavior. In our case, the rate of change of deformation is important for predicting the elastic behavior and failure of a structure. If correct, the curve contains the complete information about a physical function.

The same information about a function can be provided using equations on two levels. The basic equation gives the absolute values of coordinates for each point. This provides, however, incomplete knowledge of a function. We need a derivative equation that describes the rate of change of the function at each point. These two connected equations build the necessary logical structure of a theory. Only by proving this system of equations can we validate the theory. The linear theory of elasticity has no derivative equation for describing the variable rate of change of deformation in the interval of failure of elastic behavior. This theory is incomplete, and therefore impossible to prove. However, it can be disproved if a new theory that has complete structure can be proven. It is a rule of physical determinism that only one description of a physical phenomenon can be correct.

Formal logic requires experimental verification of inference. But inference from the basic equation is a derivative function that has a relative value of rate as its result. Such an equation cannot be verified experimentally. The relative value can be calculated but not measured and proved experimentally. It is correct if the basic equation describes the relations correctly.

This circle of logical insufficiency can be overcome if the derivative of the basic function is identified with a known function. For example, the derivative of the equation of deformation for a constant force can be identified with the tangent function. This function is selected because the character of the changes of deformations is similar to the character of the tangent function. The properties of the tangent function are known and need no confirmation. The variables in a physical function are the physical concepts. We can select these physical concepts in a way that satisfies the requirement that the derivative equation is the characteristic part of tangent function.

The first requirement, then, is that the basic equation of deformation occupies the domain of a tangent function. The domain of the tangent function can be presented in two intervals: the interval of slow changes, $0 < \tan\alpha < 1$, and the interval of rapid changes, $1 < \tan\alpha < 10$. To show that the particular physical function exists in the domain of the tangent function let us perform the numerical analysis of a function. A number can be presented as $N = a * 10^b$, where a is the significant figure $1 < a < 10$, and b is a positive or negative integer. Then, the condition for a function to occupy both intervals of the tangent function is satisfied if $\mathrm{d}Y/\mathrm{d}X$ may exist in the interval of significant figures $1 < \mathrm{d}Y/\mathrm{d}X < 10$. Then the derivative function also necessarily exists in the interval of slow changes $0 < \mathrm{d}Y/\mathrm{d}X < 1$ for it is the starting interval of the continuous function. Placing the physical function in the domain of the tangent function lets us predict the behavior of a structure at any point of the function and ultimately the failure of a structure.

Let us consider an experiment. The experiment is conducted on a number of similar specimens, i.e. rods made of the same material and with the same length. The only physical variable that distinguishes them is their diameter. The specimens are subjected to the same tensile force N. The elongation of each specimen can be measured or calculated. The standard formula for calculating elongation, $e = NL/EA$ is proven to give accurate results for e that can be considered as experimental results. The same results can be obtained with different descriptions. For example, substituting $A = \pi d^2/4$ we obtain $e = 4NL/E\pi d^2$. Substituting the ratio $A/L = R$, where R is the characteristic of geometrical stiffness, we get the formula for elongation $e = N/ER$. The results from calculating elongation for the specimen with diameter d using these different equations will be identical and true. However, these equations describe the relations between the variables differently. The mathematical inferences from these equations are different and so are the physical conclusions. In the physical world only one of these descriptions can be correct. Thus, in the standard equation of elongation, the variable d is part of the area of cross-section. The rate of change of deformation depends on the change of A described with the derivative $de/dA = -NL/EA^2$. For example, for the specimen with $A = 0.2$ in^2, $L = 20$ in, the maximum elastic force corresponding to the limit of elasticity of the material σ_y=60,000 psi is $N = \sigma_y * A = 12{,}000$ lbs, $E = 30 \times 10^6$ psi, $e = 4 \times 10^{-2}$ in, $de/dA = [2 \times 10^{-1}](\tan 11.3^\circ)$. Although we selected an elastic force corresponding to the limit of elasticity of the material, the rate of change of deformation described with the inference from the standard equation does not indicate the rapid changes characteristic of the limit.

In the equation of elongation $e = N/ER$ the variable R is an independent variable that changes depending on the diameter of the specimen. Geometrical stiffness in our example is $R = A/L = 0.2/20 = 0.01$ in, $de/dR = -N/ER^2 = [4](\tan 76^\circ)$. This rate of change of deformation comprehensively describes rapid changes. In the experiment on the number of specimens, failure was accurately predicted with the new equation of deformation and its derivative in the interval of a curve with tan 76°.

The standard equation of deformation in the domain of elastic relations with the boundaries set by the limit of elasticity of material is a linear function. It does not exist in the interval of rapid changes. There is no certain comprehensible rate that is characteristic for an elastic failure in the interval of slow changes. In order to describe physical changes correctly we have to select an independent variable in a way that places the equation of deformation in the domain of the tangent function and specifically in the interval describing rapid changes, where the prediction of a limit is possible.

The equation of deformation describes the effect of elastic force, material and geometry of a structure on deformation. The effect of geometry of a structure as a physical whole can be presented with the new complex variable defined as geometrical stiffness, in this case $R = A/L$. This variable can place the equation of deformation in the domain of the tangent function, in our example, $e = N/ER$ and $\mathrm{d}e/\mathrm{d}R = -N/ER^2$. In order to have this derivative in the interval of rapid changes, the geometrical stiffness R is selected in the interval $1 < \mathrm{d}e/\mathrm{d}R < 10$. Experimentally the interval of rapid changes for structures made of steel is $1 < \mathrm{d}e/\mathrm{d}R < 4$ (tan 76°). The geometrical stiffness that corresponds to the limit is $R_{cr} = \sqrt{N/EC_s}$, where the coefficient of elastic stability $C_s = 4$.

Let us consider another example: Tensile force $N = 6{,}000$ lbs, $L = 30$ in, $E = 30 \times 10^6$ psi, $\sigma_y = 60{,}000$ psi. Failure of a structure due to geometry may occur at the point corresponding to R_{cr}. We have

$$\frac{\mathrm{d}e}{\mathrm{d}R} = -\frac{N}{ER^2} = 4\ (\tan 76^\circ), \quad R_{cr} = \sqrt{\frac{\mathrm{N}}{4E}} = \sqrt{\frac{12{,}000}{4*30\times10^6}} = 0.01\ \text{in},$$

$$A_{cr} = R_{cr}*L = 0.01*30 = 0.3\ \text{in}^2, \quad \sigma_{cr} = \frac{N}{A_{cr}} = \frac{6{,}000}{0.3} = 20{,}000\ \text{psi}.$$

Although the material limit in this case is $\sigma_y = 60{,}000$ psi, the actual stress at failure for the specimen will be $\sigma_{cr} = 20{,}000$ psi. In this case the geometry of the specimen rather than the material forms the limitation. The actual limit of elasticity is relative. It can be the limit of the material, or the limit of a specific structure that depends on its geometrical stiffness. The lower of the two is the actual limit of the structure. The new equation of deformation and the derivative of this new equation allow us to predict the limit for the particular structure, while the standard equation of deformation does not.

According to the new equation of deformation, $e = N/ER$, the elastic deformation e is proportional to the elastic force N and inversely proportional to the resistance of the structure to deformation, ER. This resistance depends on the elasticity E of the material and on the geometrical stiffness R. The main component of geometrical stiffness in the case of tension is the ratio A/L. The physical meaning of geometrical stiffness R in this case is that it is proportional to the area of cross-section and inversely proportional to the length. In the equation of deformation, however, geometrical stiffness is present as a physical entity. Observe that

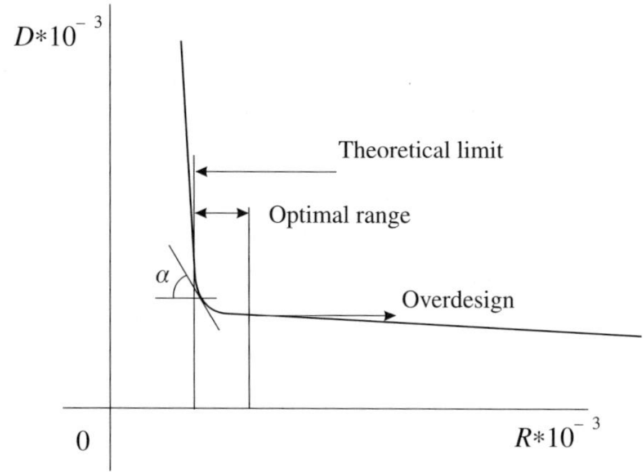

Illustration 3. Diagram deformation vs. geometrical stiffness shows the elastic behavior of a family of similar structures like in the intervals of slow changes so and in the interval of rapid changes. That allows us predict the elastic limit in the interval of rapid changes and select a structure of optimal desirable behavior using the coefficient of elastic stability $C_s = [\tan \alpha]$.

the notion of stiffness does exist in the literature. But there is no concept of geometrical stiffness as the single concept variable in the equation of deformation. This constitutes a principal difference. The derivative equation $de/dR = -N/ER^2$ describes the rate of change of deformation due to the change of geometrical stiffness. Presented graphically the function e vs. R (Illustration 3) shows changes in the interval of slow changes of its derivative and in the interval of rapid changes. The limit of this physical function can be expected in the interval of rapid changes. In the tensile test the limit of elasticity for the different specimens was predicted correctly at the point of the curve with tan 76°. The correspondence of such rate to the limit is comprehensible because small changes of the variables in the equation of deformation result in rapid increase of deformation and failure of elastic behavior.

Thus, we have built a physical function that is represented by two equations at different levels. These equations give complete and certain information about the function. Certainty is attained by identification of the derivative with the tangent function. The corresponding basic equation that operates only with the absolute values of the physical concepts can be verified experimentally in all intervals of this non-linear function. Graphical presentation of the function e vs. R completely reflects these two equations and provides an adequate description of the physical phenomenon. Note that there is no principal difference between presenting a function as a curve or in the form of the two equations, basic and derivative. The important part is that the analytical solution should correspond to the curve which is often obtained experimentally. The independent variable R that is selected to satisfy the logical structure also gives an adequate physical presentation. Geometrical stiffness has physical meaning and can be measured. In the theory of elasticity it can be considered the logical bridge between theory and phenomenon, for it is justified logically and proven to be an objective physical characteristic.

3.1 Summary

A physical function can be presented graphically as a curve and analytically in the form of two equations, basic and derivative. A curve in a plane gives three characteristics for each point: the absolute values of the coordinates (x, y) and the relative position with the tangent to the curve at this point. The same information can be obtained using the basic and derivative equations. The derivative equation has as its result a relative characteristic that describes a rate of change of a function. The absolute values of the physical concepts can be measured and verified experimentally. But it is impossible to measure relative values. These values are hypothetical unless they belong to a function whose properties are known and do not need experimental verification. In the theory of elasticity the derivative of the function is identified with the tangent function. In order for the function to have its derivative identified with the tangent function, the basic equation should exist in the domain of the tangent function and its derivative at any point should be a part of the tangent function. The tangent function can be presented in two intervals, the interval of rapid changes and the interval of slow changes. The equation that exists in the interval of rapid changes also necessarily exists in the interval of slow changes. An independent variable in the equation is selected to satisfy this logical demand. This condition is fulfilled if the derivative exists in the interval of significant figures. Thus, in constructing a purely logical–mathematical structure of the theory we are able to select physical concepts that describe physical relations adequately. At the same time the basic equation can be verified experimentally. The derivative equation also can be tested experimentally in the interval of rapid changes and thus supported with obvious physical changes. Note that in some other cases we need to change both the independent variable and other variables in the hypothetical basic equation to satisfy the condition that the derivative correspond at any point to the tangent function value. In this way the self-proved Non-Linear Theory of Elasticity has been constructed.

4. Principles of logic in NLTE

Physical theory is not simply a combination of physical ideas and mathematical descriptions of these ideas. Physical theory has to give an explanation why the particular system of mathematical formulas would necessarily yield reliable physical results. In order to have such confidence we need the logical structure of a theory that is consistent in itself, like the theorems of geometry, and at the same time has physical content that justifies the mathematical formulas and systems. Mathematical and empirical methods are insufficient for building and proving an adequate physical theory. Mathematics is an instrument for describing the initially hypothetical proposition, but has no means for judging it. Empirical methods can provide us with the facts relating to the phenomenon. However, these facts can be described and interpreted in different ways. Empirical methods have no means for judging the selection of concepts for a proposition. The only way to support a theory is to have a consistent logical structure of a theory that allows make conclusions that can be objectively explained and justified.

The engineer needs reliable methods of logical reasoning from the beginning to the end of the process of constructing and testing the scientific theory. The tools of logic are its methods for testing terms of a theory and truth of the physical relations. The main methods that classical logic proposes for reaching true conclusions are inductive and deductive. We will use these methods but only after reviewing the definitions of these methods and the rules for using them.

The method of induction is usually identified with the principle of generalization based on multiple observations and experiments. “If P has happened m times, it will probably happen $m+n$ times, the probability increasing as m increases." (A. Castell, College Logic.) We may agree then with the following conclusion – "Since the postulate of induction constitutes a mere assumption and is therefore speculative, all empirical laws are necessarily speculative." (A. D'Abro, The Rise of a new Physics.)

This definition of induction lacks the part that would describe *how* the inference from the data to conclusion is made. The process relates no more to logic than the behavior of a fox hunting rabbit relates to logic. In the argument "If m, then $m+n$", the inference, no matter how true it may be, is based more on intuition than on some constructive logical method that gives an explanation for the transition from data to conclusion. This does not mean that we should neglect such empirical assumptions in science, but until we know how intuition works it can not be considered as logical inference.

However, the constructive inductive conclusion can be referred to the inference from general laws, definitions, and generalized ideas to particular conclusions with the reference to such laws. Furthermore, if a conclusion follows from a universal law, such as the definition of tangent, then it refers to the values and character of that law (the tangent function) in all particular cases. Such a conclusion is made with an inductive procedure which we may rely on.

The other method of classical logic is deduction. Logic that is based on the argument "If P, then Q" is called deductive logic. While the inductive method in our definition relies on a priori known and tested laws and definitions in making the transition from general proposition to particular conclusion, the deductive method relies on the strict rules of deduction. In physics these are, usually, the rules of mathematics. The uncertainty and incompleteness of the method of deduction in its current form is in the fact that no reliable logical method is offered for testing proposition and/or conclusion. The deductive method is incomplete without the inductive method.

Thus we arrive at the conclusion that the deductive method may lead to provable results if a propositional function or inferential function is covered or included as a characteristic part of the universal function that mathematically describes relations similar to the observable physical relations under consideration. In NLTE one such controlling function is the tangent function. It describes the rate of change of deformation in the domain of slow and rapid changes of deformation. This reference to a universal function made the system of propositional and derivative function necessarily consistent. Theoretical laws can be used, after empirical validation, as the basis for other inductive inferences. For example, the law of elastic relations in case of static forces can be extended for dynamic forces. The methodology used for establishing the elastic relations can be used for establishing similar relations in other fields of physical engineering.

We now return to the discussion regarding the common methods of logic. The difference between inductive and deductive methods is also in the fact that inductive inference does not require empirical verification for it is assumed that such verification has been done while establishing the a priori general law. However, for a system built on the principles of deductive method it is necessary to validate experimentally/empirically the propositional function. We need the inductive method as well for including the deductive system in the body of existing knowledge.

An empirical method used for testing functions is not necessarily inductive. The empirical method gives only statistical results of verification a proposition. Mathematical statistic also allows one to calculate the necessary number of experiments for a statistically reliable confirmation of the system of equations obtained in the deductive–inductive procedure. Thus, for constructing a physical theory we use inductive and deductive logical methods of justification, and as well empirical and mathematical methods that are considered as auxiliary methods apart from logic. Further we will discuss the methods that can be reliably used in the analysis, construction and verification of NLTE in more details.

The logical steps in the process of building a theory that need to be secured are the following:

(1) Distinguishing cause from effect. Connecting cause and effect.
(2) Making a general description of the law.
(3) Selecting the terms for the description.
(4) Distinguishing empirical terms from theoretical terms.

(5) Describing in mathematical terms the possible relations between the terms.
(6) Describing the empirical domain of the phenomenon.
(7) Describing the theoretical domain of the phenomenon.
(8) Comparing the theoretical domain with the empirical domain.
(9) Establishing the logical connection between the propositional description of the phenomenon and the inference from the description.
(10) Testing the empirical terms.
(11) Testing the theoretical terms.
(12) Testing the logical structure of the theory.
(13) Distinguishing logical inductive and deductive methods from empirical and mathematical procedures.
(14) Defining inductive and deductive methods.

1. Distinguishing cause from effect. This is one of the cardinal questions in physics and in logic. The English philosopher J.S. Mill formulated the problem as follows: "The question is how to resolve this complex [cause–effect] – and to assign to each portion of the antecedent the portion of the consequent which is attendant on it."

It seems not to be that complicated a task to separate cause from effect in a real occurrence and when a theory is well developed already. Usually this connection is viewed as time-bound, i.e. first there is a cause, then there is an effect. But confusion still abounds in the theory of elasticity. Actually, the separation of cause and effect is left to intuition and is often assigned to the temporal priority. The temporal priority is important for the identification of the cause, but the cause is not necessarily evident. The conclusion that a cause exists can be inductive. If a change occurs there should be a cause for it.

The effect also, as a rule, has terms that are obtained with inductive process. Thus, external force applied to a body has temporal priority to the following deformation. But also it should be more clearly realized that the external force that is the cause of the deformation is not a term in the law that describes the elastic relations in a body. In the equation of deformation the force term relates to the elastic force. The cause (external force) is independent of the effect, namely it is independent of the body and consequently from the deformed body. Therefore an external force cannot be part of the function describing the relations in a body. On the other hand, the elastic force is not the cause of the deformation but part of the effect described as elastic relations in a body. The difference between the two forces is more clearly understood by following the logical procedure and distinguishing cause from the effect. The external force is, in principle, observable. It can be seen as an entity separate from the structure in question and it can be measured even before it is applied to the body. On the other hand, the elastic force is an unobservable theoretical term, which cannot be measured but can be calculated. How then are cause and effect connected?

2. Making a general description of the law. The effect (in this case, the deformation of a body) depends on the external force that is the cause. More correctly, it depends on the work performed by the external force. The logical relation between cause and effect exists, but it is a non-functional relation. We may call such relation an inductive relation that is validated by the inductive method of logic. We can say that the relation of the elastic phenomenon to the outside force obeys a general law that is adopted by science. Such a general law that would connect cause and effect in the theory of elasticity is the law of conservation of energy. Work of an external force acting on a body is stored as an elastic potential energy of that body. We can calculate the work and equate it to the elastic energy. Here the method is called inductive. In making inductive inference we rely on the knowledge of some general laws and definitions rather than on the rules of mathematical inferences. The inductive method acts under the umbrella of an applicable general law. In that capacity it is as definitive and reliable justification for inference as the deductive method unless the scientific system is rejected. The reliability of the inductive method depends on the system of knowledge in which we operate. The inductive method further leads to the formulation of a propositional function, and then deduction leads to the conclusion from the proposition.

The description of the phenomenon of elasticity is based on the idea that exists in our system of knowledge. More specific, in classical mechanics we accepted the definition that if work is performed, then there should be a force that does the work. After removing the external force the body returns to its initial form and position. It means that reverse work is done in the absence of a visible force. The conclusion is made that there is an unobservable elastic force that performs this work. The proposition based on this idea is Hooke's law, which states that in an infinitesimal volume of a body the unit elastic force/stress is proportional to unit deformation/strain, $\sigma = E\varepsilon$, where σ is the unobservable elastic force and ε is the observable measurable deformation. Here again the reliable basis of the particular law is the general idea of a force that is connected with the definition of work rather than the observations that lead to the generalization. Can we devise a theory that does not involve unobservable entities? Yes, that is possible. But the concept of elastic force, though unobservable, is still very convenient and gives reliable practical results. It also connects cause and effect.

3. Selecting the terms for the description. The description of the elastic relations in the deformed body in NLTE for a whole body is $D = F/ER$: the deformation is proportional to the elastic force and inversely proportional to the resistance of a body to deformation. From the logical point of view the terms we selected for describing the law need to be separated into observable terms that can be measured and unobservable terms that cannot be measured but obtained with inductive method. Need for proof arises with the construction of a propositional function. All terms in the function are selected on the basis of their relations to the function. The deformation D is a function that can be measured. The magnitude of an observable term is independent from judgment as to true or false. It is true in itself, but it might be wrongly selected in the relation to the function. Logical judgment refers to the relations. The geometrical stiffness R designates the resistance of a structure to deformation due to the geometry of the structure; it is a compound observable term. The elastic force F is an unobservable term obtained with the inductive process. It is proportional to deformation according to Hooke's general law. The modulus of elasticity E of the material is a theoretically established ratio: $E = \sigma/\varepsilon$. As we can see these terms of the description are different in their relation to reality. Some of the terms, such as deformation D and geometrical stiffness R, are observable and measurable directly or indirectly. Other terms, such as elastic force F and coefficient E, are theoretical terms; they are non-observable and their values can be calculated but not measured. A conclusion based on the general law of energy conservation is accepted in science and we can make reliable conclusions based on this law. The fact that there was a time when this so-called 'universal law' needed validation does not undermine its existential validity. After the law of elasticity would be proved it also will be a universal law in this system. The law of conservation of energy allows one to calculate the work performed by the 'elastic force' and thus to calculate the elastic force. Elastic deformation is a reversible deformation, therefore the elastic force is equated to the external force that causes deformation. The external force is measurable. It is a matter of convenience to obtain the magnitude of the elastic force as the arithmetic mean value of the external force applied to a body. In order not to miss some external force we can check the correspondence of the calculated value of deformation with the real deformation.

However, identification of the elastic force in the function of deformation with the external force is not justified from the logical point of view. The elastic force according to the accepted Hooke's law is proportional to deformation. Deformation is distributed throughout the material of a structure so should be true for elastic force. Its distribution is connected with the geometry of a body rather than with the distribution of external force. Even in such cases, for example, when concentrated force is applied to a thin shell and causes distribution of elastic deformation and force pertaining to the area close to the application of external force such distribution is also due to the design of a structure.

4. Distinguishing empirical terms from theoretical terms. The terms for the description of elastic relations can be selected in different ways. Thus the traditional description of elongation of a bar is $e = NL/EA$. In the non-linear theory of elasticity the description is $e = N/ER$. Mathematically all descriptions are true, and descriptions that give the same result for a function are defined as equivalent descriptions. But in logic different descriptions have different terms and different relations between the terms and subsequently a different logical structure of the system describing the phenomenon. Only one of all possible systems can be logically proven. We need a rule for selecting the justifiable description.

5. Describing in mathematical terms the possible relations between the terms. The empirical verification of a propositional function does not provide logical proof for the system. For example, consider $e = NL/EA$

and $e = N/ER$, where $R = A/L$. These two descriptions give the same result for the measurable elongation e of a bar. The geometrical terms in both equations are measurable data that we have no reason to doubt. The elastic force is a constant in both equations and has the same theoretical value that we have no reason to doubt. The same work, the same deformation and thus the same elastic force. However these equations are two different bases for different mathematical and logical inferences, only one of which can be logically proven. And the one that has logical justification is also existentially correct. The partial derivative for one equation is $de/dR = -N/ER^2 = -NL^2/EA^2$; for the other equation it is $de/dA = -NL/EA^2$. These equations describe the rate of change of deformation differently for the same data: N, A, L, E. However it cannot be that both conclusions are true. We need proof for the selected description. And here proof is in the hand of logic.

6. Describing the empirical domain of a phenomenon. There are some conditions to be satisfied in order to prove a theory. Thus, the proof of a proposition is possible only if a function has limits, or, in other words, if the domain of a function is established. There are two types of limits of elastic relations; one is theoretical, based on the description of elastic relations; the other stems from empirical data that may or may not support the theoretically obtained limit. The empirical data may challenge the theoretical conclusion regarding the limits. The equation of deformation in any of its forms can be written as $y = k/x$. This type of equation has a mathematical limit. In physics we are not bounded by a mathematical limit. The reason for that is that the mathematical limit does not belong to the function. The mathematical limit of a continuous non-linear function is outside of the function, while the limit we are looking for has to be a part of the function and part of the selected physical set for an independent variable $R(x)$ or $A(x)$. The theoretical limit for the set is described by a tangent function in the interval of rapid changes. The individual limit is a product of inductive conclusion. First, it is connected to the universal function that is a tangent. It is also connected to the digital numerical system, $N = a * 10^b$, or for that matter to another numerical system. The magnitude of rate of deformation should be in the interval for significant figures (a), $0 < \tan \alpha < 10$. Only this interval has physical significance because comparison of magnitudes for a physical concept is important at its own level.

The other condition for the selection of a limit is that the individual limit should not exceed the limit of the material. Comparison of these two values will accomplish the selection.

7. Describing the theoretical domain of the function. The conclusion as to existence of a theoretical limit is made based on the character of the mathematical relations in the propositional function. It would be an impossible task to obtain the individual theoretical limit a priori to establishing the logical structure. Although there are observations that show that the limit may depend on the geometry of a structure, certainty comes only with a mathematical analysis of the propositional function that indicates a mathematical limit. Such a conclusion is inductive for it is based on the knowledge of the character of the function rather than on the observational clues.

8. Comparing the theoretical domain with the empirical domain. The empirical confirmation of the elastic limit, however, presents a challenge. For some specimens the limit has the theoretical calculated value. In other cases, when the theoretical limit is higher than the limit of the material, failure occurs at the limit of material. The relative character of the limit could be explained if there was a universal law for the relation between a geometrical whole and the material parts of the whole. There is yet no such law. Therefore in NLTE the relative character of the limit of elasticity is accepted as fact. A multitude of experiments performed on the standard specimen shows that the limit depends on the material. On the other hand the logic of the mathematical structure of the law of elasticity shows that there is a limit depending on the geometrical stiffness of a structure. It is two independent factors that establish maximum value of elastic force within the elastic limit.

9. Establishing the logical connection between the propositional description of the phenomenon and the inference from the description. Not every step of the reasoning should be attributed to deductive or inductive logic, but only those that require proof. A comparison of natural numbers does not require proof, at least not in physics. The hypothesis that can be considered as an explanation of the relative character of the limit of elasticity is as follows. The parts of a whole have their own geometrical structure and therefore their own limits. Also, the whole body can be considered as a part that has its own geometry and strength characteristics. The characteristic of the whole can be compared with the characteristic of the parts. For a hypothesis to become a universal law similar patterns of behavior need to be noticed for the other phenomena. There is a principle of uniformity in our views about the world. I believe that the relation between a whole and its parts merits further investigation.

In physics the proposition of an argument has the form of a propositional function. The propositional function describes relations between the terms that are characteristic for the phenomenon. From a logical/physical pair cause–effect the propositional description refers to the 'effect' only. The reason for that is that cause (external force) is an entity that is independent from effect (what happens to the body). No functional connection can be established between cause and effect while a proposition in physics is a function.

10/11. Testing the empirical and theoretical terms. The terms in a function are of two types, i.e. observable and unobservable or theoretical. An observable term can be measured. There is no need to prove its existential nature. Nevertheless it is necessary to explain the selection both of observable and of unobservable terms. An unobservable term, on the other hand, requires inductive validation for its existence. This can be done under the umbrella of a universal law. An unobservable term cannot be measured but it can be calculated using the equation provided with inductive reasoning and a universal law.

The propositional function that was constructed need validation because there is usually a number of ways to build such a function depending on the selection of terms. Empirical or experimental verification may support the function, but that is not yet confirmation of its existential validity. We might say that the propositional function is valid if it is part of a valid logical structure *and* it is supported by the empirical/experimental data. These are the two conditions necessary for the proof of a proposition.

12. Testing the logical structure of a theory. What kind of logical structure can be considered to be valid? A logical argument has two parts, i.e., a propositional function and the inference from this propositional function. The inference is obtained with the deductive method. In physics the propositional function is a basic physical function; by applying mathematical rules we obtain the derivative function that describes the relation between the function and the selected independent variable.

The proof or falsification of the system of this pair of equations first comes with the connection of the inferential equation with some universal pattern describing similar relations. The function of deformation has a limit. This limit is characterized by a rapid increase in the rate of change of deformation. The rate is described with the tangent to the curve representing the propositional function. This does not yet mean that the inferential function should always be identified with a tangent function, for the character of the derivative function depends on the character of the basic function. However in the case of elastic deformation induced by a constant force, the character of a smooth hyperbolic function in two intervals of the tangent function implies that the tangent function is a derivative of the equation of elastic deformation. Or rather, it is a part of the tangent function in the interval $0 < f'(x) < 4$.

The identification of the inferential function with the universal tangent function is necessarily an inductive conclusion. It allows us to make a proper selection of terms and, if necessary, to correct the propositional function accordingly. This method of constructing a verifiable logical system is brought about by the combination of deductive and inductive methods. None of these two methods by itself is sufficient for validating a theory.

13. Distinguishing logical inductive and deductive methods from empirical and mathematical procedures. The structure of a physical theory is not purely mathematical. Although a physical theory has mathematically constructed parts and uses mathematical procedures, it also employs inductive and deductive means for constructing and justifying the logical frame, and empirical means for testing it.

A propositional function differs from a purely mathematical function. Any purely mathematical function is a final construction. A propositional function in a logical argument is only a part of the argument. It can be changed in the further process of logical reasoning. This is not characteristic of a mathematical function that does not require justification.

A propositional function has terms of different origins with different foundations, i.e. physical empirical and inductive theoretical. In mathematics such division has no significance. In physics, however, it makes a difference, and employment of logic is needed for dealing with the theoretical terms.

The empirical terms such as deformation *D* and geometrical stiffness *R* are observable and measurable. Geometrical stiffness can be presented as a compound logical statement but in the propositional function it is a whole physical concept. It is an important logical demand to treat the terms of mathematical equations in physics as physical entities.

The observable terms do not require logical proof for establishing their existence and values. The theoretical terms in the proposition, such as elastic force *F* and modulus of elasticity *E* need logical proof for establishing their necessity and their quantitative values. This proof can be found in the general laws that are accepted as a priori to the particular propositional function. A priori here does not necessarily mean that we forget how these principles were obtained initially. It does not mean that the terms are completely abstracted of physical meaning. Here, at least, the notion 'a priori' is used for laws that were formulated prior to the scientific discipline we are dealing with. These laws were formulated prior to using them in the particular theory and they are more general in character than the laws in question. Thus, the law of conservation of mass–energy has earned its place as a general law that can be used as a priori knowledge that we do not question here.

The theoretical meaning of elastic force can be explained from the point of view of the law of conservation of elastic energy. Such compliance of a theoretical term with some general law is a necessary inductive procedure that gives a measure of reliability to the proposition. It is worth mentioning that at this moment in time we are not in the Stone Age, using only superficial observations for our conclusions. We are not in the Greek's speculative period, when propositions were accepted without any proof. We have reliable knowledge that can be used as a priori knowledge. The accuracy and reliability of the deformation calculated with the propositional function lies in its correspondence with experimental result. This correspondence also supports the selection of the theoretical terms.

However, the same results can be reached with different selections of the empirical terms, as was discussed already. For proving logically the existential value of the propositional function we first employ the method of deduction. The second part of an argument the conclusion, which is a function obtained as mathematical inference. The deductive method is presently identified with such inference. Actually, without further logical analysis such inference is not proved to be a requirement for the physical theory.

Then the next step of the logical analysis should be the comparison of this deductive derivative with the universal tangent function and with the dynamics of the actual behavior of the function in its physical domain. The function of deformation has a limit. Mathematical analysis of the function allows one to establish the limit. For example, the analysis of a function of type $y = k/x$ shows that it is a hyperbola. The physical limit can be selected with consideration of the elusive mathematical limit. Nevertheless it is not the mathematical limit. The mathematical limit does not belong to the function, it is outside it. In physics we are interested in the limit that belongs to the physical set described with the function.

The limit of the set is selected in the interval of rapid changes of the derivative that is a tangent function. In order for us to do such analysis the propositional function in question should exist in the interval of rapid changes of the tangent function. The explanation for this is an inductive conclusion from the most general idea that the property of one set under consideration has to be part of another similar set that is not in question but is a characteristic part of the universal function. This conclusion leads to the selection of an independent variable and building the function. Then the mathematical conclusion for establishing the domain is $0 < f'(x) < 4$, where $f'(x)$ is identified with tangent in this interval and has terms that should be terms of $f(x)$. Experimental results obtained on a number of specimens will help in selecting the concrete limit. Should we call this selection inductive because it rests on some observations?

Thousands of years of observation of the elastic behavior of solid bodies do not lead to the law of elasticity. Inductive conclusions in all cases when we do such conclusions are based on some already formulated and accepted general idea or general law that may lead further to the mathematical description. These descriptions are as reliable as inference obtained mathematically. The difference, however, is that the mathematical deductive conclusion needs to be tested on its corroboration with the facts, while terms and laws obtained with the inductive method do not need such confirmation. The deductive connection of a propositional function and an inferential function is also proven to be true if the inferential function is a characteristic part of a universal function. The system created in such manner is necessarily logically correct. The tangent function became the logical pattern forming the function in question. Empirical validation of this logical system makes a law out of this system.

Is it always the case that the tangent function is a pattern of behavior forming a function? We may consider a function for which the deformation and correspondingly the elastic force are changing due to the dynamic character of the external force acting on the structure. Depending on the law of change of deformation for the dynamic force we may obtain by deductive inference a law for the rate of change of deformation that is characteristic for each structure. For example, for the individual structure if the dynamic force changes according to a sine, then its derivative describing the rate of change forms a cosine. However the changes for a set of structures remain to be described with tangent function. It is important to note that the limit for the rate is the same as the limit in case of constant force, i.e. $dy/dx < 4$ ($\tan 76^{\circ}$). The interval $0 < f'(x) < 4$ is the domain for the function of deformation for any type of loading. The established limit for the rate of change of deformation is a constant criterion. This value does not depend on elastic force or deformation. However backward decision namely selection of the limiting deformation and stress that is based on this constant criterion of elastic stability is the necessary logical condition.

14. Defining inductive and deductive methods. Two methods of logical reasoning leading to reliable conclusions were employed in NLTE. They are used whenever there are appropriate conditions that these methods can resolve. The methods are distinct with respect to their means for reaching conclusions. (1) The deductive method in physics is based on inference with strict mathematical rules, such as from propositional function to derivative function. (2) The inductive method is used for connecting the theoretical concepts and particular laws with universal laws and definitions, and thus to justify these particulars. It is possible to make a reliable inference from general to particular. It is impossible to make a logical inference from particular to the general. The relation between general and particular is asymmetric.

In NLTE, the inductive method is used for conceiving the notion 'elastic force' and for devising the equations connecting elastic force and deformation for infinitesimal element. Also, the inductive method is used for placing the function of deformation in the domain of the tangent function. The inductive conclusion then is that the derivative function should exist in the interval of rapid changes. Further it allows one to conclude that in order to satisfy the requirement $1 < f'(x) < 10$ the order of magnitude of an independent variable $a * 10^b$ should be the same as for deformation $a_1 * 10^b$. Otherwise the set of the function appears at a different level than the set of the independent variable, which would make comparison of the sets impossible. It is necessary to do all comparisons at the same level of magnitude. Consequently, the inductive method allows selection of an independent variable using the mathematical analysis.

The inductive method can be defined as inference from general laws to particular laws. This understanding makes the inductive method as definitive as the deductive method. The inductive method connects the unobservable terms, the mathematical conjunction of proposition and conclusion, and cause and effect to general laws and definitions. We may say that the deductive method is a method of logical transition from one statement to another. The combination of deductive and inductive methods lets us build a reliable logical structure of a theory.

There is also empirical method that is used for validation of observable properties of phenomenon. The empirical method is different from inductive method of reasoning. The empirical method provides empirical data rather than means of reasoning. The empirical data by itself does not lead to a conclusion. Although it is mentioned, probably, in every book on logic, nevertheless the confusion in distinguishing the empirical method of testing from the inductive method of logic remains. A structured experiment requires a purposeful idea already. And there is no need for justification of experimental data.

Logic provides true or falsehood judgment to the logical system that is the framework of a physical theory.

5. Method of optimal structural design

Historically the method of structural design has been based on the Linear Theory of Infinitesimal Deformations. This theory to some extent solves the problems of determining deformations and stresses. However, it provides no methods for analyzing deformations, predicting the individual limit of a structure, comparing the elastic properties of the structure, or for optimizing the structure.

This section describes a method of optimizating a structure by approaching the structure as a geometrical whole. The method is based on the principles of the Non-Linear Theory of Elasticity described above. It allows us to obtain the individual limit of elasticity of a structure by using a simple non-destructive test and calculations. The method lets us compare the elastic properties and behavior of similar and different structures. The new criterion of elastic stability makes the calculation of the optimal dimensions a uniformly reliable process.

This method of optimal structural design is based on the introduction of a new physical concept of strength. Currently the limit of elasticity of the material is the general criterion for structural design. Experimental limiting deformation sometimes is added for the evaluation of structures. According to the new point of view each structure has an individual elastic limit which, in general, is different from the limit of a material. The limiting deformation corresponds to the individual elastic limit. The individual limit for a structure depends on the resistance of a structure to elastic deformation due to the stiffness of the structure.

The stiffness of a structure depends on the elastic properties of the material, the geometry of the structure (dimensions, shape, specifics of design, boundary conditions, or supports) and the mode of deformation. The stiffness of a structure, depending on the geometry and the boundary conditions, is separated and described as the geometrical stiffness of a structure. The equation of elastic deformation in its general form, then, can be written as follows. The total deformation of a structure is proportional to the mean value of the distributed elastic force and inversely proportional to the modulus of elasticity of the material and the geometrical stiffness, $D = F/ER$. Thus, in the case of tension, following the equations of linear theory, $e = N/ER$ where $R = KA/L$. In the case of bending, $\theta = M/ER$ where $R = KI/L$. The coefficient K reflects the effect of design specifics on the geometrical stiffness.

The physical meaning of geometrical stiffness can be explained with the following examples:

(1) Two beams of the same material elasticity E may have different resistance to deformation if they have different geometrical stiffness.

(2) Two similar beams with equal moments of inertia of the cross-section I_x may have different geometrical stiffness if the lengths of the beams are different. The longer the beam, the less its geometrical stiffness.

Note that in the linear theory of elasticity, structures with identical cross-sectional characteristics when subjected to equal load have the same stresses. If allowable stress is a criterion for a reliable design then the length of a structure is not taken into account. This is against even common sense. If all other conditions are the same the longer a structure the less its resistance to deformation. The concept of geometrical stiffness is necessary for the proper evaluation of a structure.

(3) Two similar beams may have different geometrical stiffness if they have the same length but different moments of inertia of their cross-sections. The geometrical stiffness is proportional to the moment of inertia of the cross-section. The larger the cross-section, the more geometrical stiffness.

(4) If two similar beams have the same ratio of the moment of inertia to the length, then these beams have the same geometrical stiffness.

(5) The geometrical stiffness of different structures can be the same if the structures are made of the same material and for the same distributed elastic moment/force have the same resulting deformation, $R = M/E\theta$. Note that the geometrical stiffness of the real structure does not depend on the elastic force. The greater the elastic force, the greater the deformation. Geometrical stiffness is a geometrical characteristic.

All of these features of geometrical stiffness are useful in the investigation of the real structures. Geometrical stiffness has units. In the case of tension/compression the unit is $R = A/L$ (in). In the case of bending/torsion it is $R = I/L$ (in^3). In some simple cases the geometrical stiffness of a structure can be calculated and compared with the geometrical stiffness of an optimal structure. However, it is often not practical to make a mathematical description of the effect of the supports and the specifics of the geometrical stiffness. Even the effect of the geometry, except in simple cases, is difficult to determine with calculations. The geometrical stiffness of a structure can be calculated or measured if the elastic force and deformation are known, $R = F/ED$. Furthermore, we can use geometrical stiffness to study typical boundary conditions and the specifics $K = RL/A$ or $K = RL/I_x$. The coefficient K is defined as the 'coefficient of specifics'. Likewise, the moment of inertia of the cross-section or section modulus can be determined if geometrical stiffness and coefficient of specifics are known. The optimal cross-sectional dimensions of a structure can be calculated if the geometrical stiffness corresponding to the limit of elasticity of the structure is known.

The limit of elasticity is characterized by a rapid increase of deformation. The rate of change of deformation can be described with a derivative equation. For tension it is $\mathrm{d}e/\mathrm{d}R = -N/ER^2$; for bending, $\mathrm{d}\theta/\mathrm{d}R = -M/ER^2$; and in general form, $\mathrm{d}D/\mathrm{d}R = -F/ER^2$. The geometrical interpretation of the rate is that it is the tangent to the curve D vs. R at the particular point.

Let us consider a particular case: In the tensile test on a specimen with $A = 0.2\ \mathrm{in}^2$, $L = 20\ \mathrm{in}$, $E = 30 \times 10^6$ psi, elastic failure occurred for $N = 12{,}000$ lbs, $\sigma = 60{,}000$ psi, $R = 0.2/20 = 0.01$ in. From the derivative equation for tension, $\mathrm{d}e/\mathrm{d}R = -N/ER^2 = -\tan\alpha$, the rate at the point of failure can be calculated, $\tan\alpha = 12{,}000/(30 \times 10^6 * 1 \times 10^{-4}) = 4$ (tan76°). The coefficient that corresponds to the value of the tangent is defined as the coefficient of elastic stability, $C_s = -\tan\alpha$; at the point of elastic failure, $C_s = 4$. This coefficient is given in units which depend on the variables in the equation of elastic stability.

The individual limit of elasticity for a structure can be calculated using the coefficient of elastic stability. In a test on different specimens made of the same material, the limit of elasticity was properly determined until the limit of the elasticity of the material was exceeded. Elastic failure then occurred at the point of the limit of elasticity of the material. The experiments suggest that the limit of elasticity for a structure is of comparatively lesser value.

The elastic behavior of a structure is illustrated in the diagram D vs. R (Illustration 3, above). The diagram shows a rapid increase of deformation in the interval of the proportional-elastic limit. Beyond this interval, an insignificant decrease of geometrical stiffness results in failure of the elastic behavior. Increasing geometrical stiffness above the proportional limit does not improve elastic stability but does increase the weight of a structure. The rate of change of deformation is the reliable criterion for calculating safe optimal dimensions of structures. This coefficient also provides the necessary degree of similar behavior for designing similar structures. Knowledge of elastic behavior gives the designer an opportunity to select a coefficient of elastic stability with some safety factor and to calculate the optimal dimensions of similar structures. For example, if $C_s = 1$ (tan 45°) then a structure with $N = 12{,}000$ lbs, $E = 30 \times 10^6$ psi has optimal geometrical stiffness, $R_o = \sqrt{12 \times 10^3/30 \times 10^6} = 0.02$ in. If the length is $L = 20$ in, then the safe optimal cross-section is $A_o = R_o * L = 0.4\ \text{in}^2$.

Let us consider the example of a simple center-loaded beam: $P = 10{,}000$ lbs (4,536 kg), $L = 240$ in (20 ft, 6.1 m); AISI 1035 steel, $\sigma_y = 80{,}000$ psi, $E = 29 \times 10^6$ psi, mean value of elastic moment $M_m =$ PL/8 = 300,000 lbs-in. The optimal geometrical stiffness is $R_o = \sqrt{M/EC_s}$. Selecting $C_s = 2\ \text{in}^{-3}$, then $R_o = 0.07\ \text{in}^3$. The optimal moment of inertia is $I_o = R_o * L = 17\ \text{in}^4$. From among American Standard Beams sizes the closest size is 6*3, $I_x = 21.8\ \text{in}^4$, $S_x = 7.3\ \text{in}^3$, stress $\sigma = M/S_x = 41{,}096$ psi, weight per foot $w = 12.5$ lbs, total weight $W = 250$ lbs.

Calculation with the current standard method is as follows. Recommended allowable stress $\sigma_a = 20{,}000$ psi, $M_{max} =$ PL/4 = 600,000 lbs-in. Section modulus is $S_x = M_{max}/\sigma_a = 30\ \text{in}^3$. We can select the standard 12*5 beam, $S_x = 36\ \text{in}^3$, weight $w = 31.8$ lb/f, total weight $W = 636$ lbs.

The beam recommended by current standards is 2.5 times overweight in comparison with the optimal obtained with the new method. Material can be used much more economically. However, this is impossible without knowledge of the individual limit for a structure and the criterion of optimization of the structure offered by the method of optimal design. Note that optimization of a beam is calculated based on the assumption that the existing formula $\theta = (M/E) * (L/I_x)$ is correct. An alternative formula could be $D = (M/E) * (L/S_x)$. The optimal (minimal reliable) dimensions for any type of a structure, e.g. beams, plates, shells, cylinders, shafts, etc., can be calculated using this method of optimization.

The following are the common steps recommended for the optimization of a structure and a series of similar structures. The method includes a non-destructive test of the model or prototype of the structure.

(1) By applying a known load that is characteristic for a particular case, but not necessarily the actual projected load, and measuring total deformation of the structure we will determine the actual geometrical stiffness of the structure; for example, in the case of bending $R_a = M_{exp.}/ED$. Note that the geometrical stiffness of a structure is a geometrical characteristic of that structure. It does not depend on the elastic force applied to the structure.

(2) Optimal geometrical stiffness, on the other hand, depends on the elastic force in the structure. If the force is known, then optimal geometrical stiffness can be calculated using the anticipated coefficient of elastic stability C_s: $R_o = \sqrt{M_a/EC_s}$.

(3) The optimal area of cross-section, or optimal moment of inertia, or optimal section modulus can be calculated, $A_o = A_a * R_o/R_a$; $I_o = I_a * R_o/R_a$; $S_o = S_a * R_o/R_a$.

(4) The maximum stress in a structure of optimal dimensions must be compared with the elastic limit of the material, $\sigma_{max} \leqslant [\sigma_m]$.

(5) This method also allows us to calculate the actual elastic force that is distributed in a structure. By measuring deformation from the actual forces and knowing geometrical stiffness (step 1), the elastic force is $F_a = D_a E R_a$.

(6) By using the optimal geometrical stiffness for the evaluated structure we can obtain the optimal dimensions for a series of similar structures, $R_o^1/R_o = (A_0^1/A_o)(L/L^1)$, $A_0^1 = A_0(R_0^1/R_0)(L^1/L)$.

5.1 Summary

The new method of design is based on the recognition of the individual limit of elasticity of a structure. The rate of change of deformation is a good indicator of elastic behavior. A new criterion for optimal design is the coefficient of elastic stability. A new property of a structure, geometrical stiffness as a single characteristic, is introduced in order to reflect the effect of geometry on elastic behavior properly. A new equation, deformation equals elastic force divided by geometrical stiffness and modulus of elasticity of material, describes these relations. The derivative equation of elastic stability describes the relationship between elastic force and geometrical stiffness. The equations describing geometrical stiffness make it possible to compare similar and even different structures.

5.2 Example of beam design

$P = 6{,}000$ lbs; $L = 120$ in; $E = 30 \times 10^6$ psi; $M_m = PL/8$

$R \times 10^{-2}$	2	3	4	5	6	7	8	9	10
$\theta \times 10^{-2}$	15	10	7.5	6.0	5.0	4.3	3.7	3.3	3.0
$\tan\alpha = d\theta/dR$	7.5	3.3	1.9	1.2	0.8	0.6	0.5	0.4	0.3
α (°)	82	73	62	50	40	31	25	20	17
Y_{max}	3.0	2.0	1.5	1.2	1.0	0.9	0.75	0.7	0.6

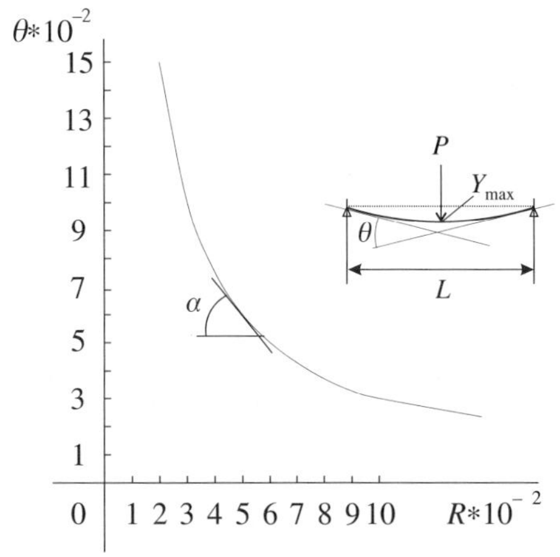

Deformation θ vs. Geometrical stiffness R

The curve shows the behavior of a family of beams for given elastic moment M and the length L for the different moment of inertia of cross-section. Comparison of rate of change of deformation $\tan\alpha$ for the structures having different geometrical stiffness R allows us to select the beam of optimal geometrical stiffness and of optimal cross-section, $I = R * L$.

6. Optimal structural design (examples)

6.1 Tension/compression and bending

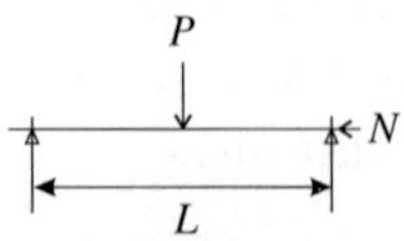

Total deformation: $D = M/ER_b + N/ER_t$
Equivalent moment: $M_e = D * ER_b$
Optimal geometrical stiffness: $R_o = (M_e/EC_s)$
Coefficient of elastic stability: $C_s = [\tan\alpha]$ ($1/\text{in}^3$)
Optimal moment of inertia of the beam: $I_o = I_b * R_o/R_b$.

6.2 Beams with multiple supports

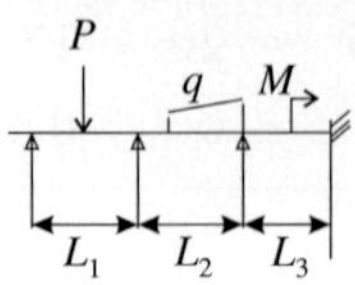

By measuring deformation and geometrical stiffness of a span we can find the moment distributed in the span, $M_1 = D_1 ER_1$.
Optimal geometrical stiffness: $R_{1,o} = \sqrt{M_1/EC_s}$
Optimal moment of inertia: $I_o = I_1 R_o/R_1$

6.3 Deformation of plates

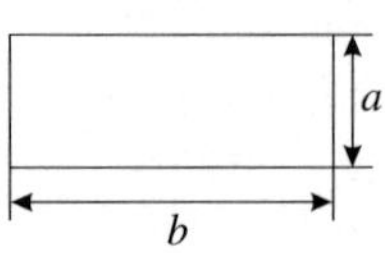

The force P is perpendicular to the surface. By measuring deformation in the direction a the actual geometrical stiffness can be determined, $R_a = M/DE$, where $M = Pa/8$ and the geometrical stiffness is $R_a = K * bt^2/6a$, where t is the thickness of the plate. Optimal geometrical stiffness $R_o = \sqrt{M/EC_S}$.
Optimal thickness of the plate is $t_o = t_a\sqrt{R_o/R_a}$.
The process can be repeated by measuring deformation in the direction b.

For any type of a structure, optimal geometrical stiffness can be calculated with the equation of elastic stability, $R_o = \sqrt{F/EC_s}$.
The description of geometrical stiffness can be verified experimentally.

7. Optimal simple beam

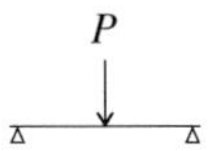

(1) $M = M_{max}/2 = PL/8$
(2) $R_o = \sqrt{M/EC_s}$, $C_s = 2$, $E = 30*10^6$
(3) $I = RL$, S is correspondingly selected from the tables of standard structural sizes.
(4) $\sigma = M/S < [\sigma_y]$

L (in) \ P (lbs)		100	300	500	800	1000
10	M		375	625		
	R		0.002	0.003		
	I		0.02	0.03		
	S		0.03	0.04		
	$\sigma*10^3$		12.5	15.6		
20	M	250	750	1,250	2,000	
	R	0.002	0.003	0.005	0.006	
	I	0.04	0.07	0.10	0.10	
	S	0.06	0.09	0.10	0.10	
	$\sigma*10^3$	4.17	8.33	12.50	20.00	
30	M	375	1,125	1,875	3,000	3,750
	R	0.002	0.004	0.006	0.007	0.008
	I	0.06	0.10	0.20	0.20	0.20
	S	0.07	0.10	0.10	0.20	0.20
	$\sigma*10^3$	5.36	8.65	13.39	15.79	19.74
40	M		1,500	2,500	4,000	5,000
	R		0.005	0.007	0.008	0.009
	I		0.20	0.30	0.30	0.40
	S		0.20	0.20	0.30	0.30
	$\sigma*10^3$		7.9	13.16	16.00	16.67

The method of optimization of structures that is based on rate of change of deformation criterion is an exclusive method for predicting the stresses for the structures with the similar elastic behavior.

8. On mathematics in physics

> A truly realistic mathematics should be conceived, in line with physics, as a branch of the theoretical construction of the one real world, and should adapt the same sober and cautious attitude towards hypothetical extensions of its foundations, as is exhibited by physics. *(H. Weyl)*

We may consider mathematical methods in physics as a set of scientific tools. We select and use these tools to solve some physical problems and deduce consequences. Although this practical approach to mathematics is understandable, especially in application to engineering problems, it is not a common approach. Mathematics was considered independent from physics. A commonly accepted point of view is that misconceptions in physics can be explained by insufficiency of the empirical data, or are due to wrong physical ideas, but false ideas in physics are never charged to uncertainty, inconsistency, or the formal use of mathematical methods. Here, in examples from the theory of elasticity, we will see that such misconceptions, generated by the formal use of the general mathematical rules, indeed, are commonplace in physics.

Mathematical descriptions often imply physical ideas. Thus, description with a linear function suggests that the function does not have a limit that depends on the variables in the function. The limit, such as the limit of elasticity of the material, is then found outside of the described relations. On the other hand, a non-linear description suggests that a limit is generated by the geometry of a structure as the culmination of the described relations. Because of this connection between mathematical description and physical implication, the formal use of mathematics in physics is not justified.

The rules of mathematics are not always in agreement with the demands of physics. Physical determinism places restrictions on the general rules of mathematics. The general demands of physics are quite understandable. A physical function and all its variables have physical meaning; the relations among the variables are specific, unique and definitive. Each variable in a physical function is a physical concept. A physical function establishes relations among some concrete physical concepts but not others. With this approach the common algebraic transformations, when applied to a physical function, change the variable concepts in a function and the relations among the concepts. The mathematical inferences, and consequently the physical inferences, are different for the function after its transformation. Even what was physically correct before transformation may become incorrect after transformation. Thus, with these requirements in mind we see that the general rule of mathematics that "a function of a function can be presented as a single function" is not applicable to the physical function because this rule allows for formal transformation and may lead to physical errors and misconceptions.

A physical function requires verification. Therefore, we construct a physical function with the purpose of creating an opportunity to test the function. This requirement sets demands on the mathematical structure of physical theory. For verification of a physical function we need the basic and derivative equations. Mathematics allows us to obtain a derivative corresponding to the basic equation regardless of whether that basic equation is true or false. In physics there is no sense in obtaining a derivative from a hypothetical equation because such inference will also be hypothetical.

We have no means to test the derivative equation that describes relations. The necessity to test physical function changes the logic of the mathematical relations. We need mathematical means to construct a basic function corresponding to the known derivative function. The rules of mathematics should be adjusted accordingly to reach this goal. Let us first consider the traditional use of mathematics in physics.

The role of mathematics has increased since the 17th century with the development of differential calculus. The reason, according to A. d'Abro in vol. 1 of "The Rise of the New Physics" is that "the mathematical expression of the relation, and hence of the law, is given by differential equation." More rapid development of the physical sciences is attributed to the employment of mathematical differential inferences. "Thanks to the mathematical instrument, formerly unsuspected relations were revealed and additional laws were derived", d'Abro wrote. Significant formalization of physics has occurred in the 20th century with the development of computer-aided methods of calculation.

As it was a century ago, the role of mathematics in physics is viewed as so obvious that the relationship between these two scientific disciplines is not discussed seriously. "So far the question is purely one to settle between them, on the basis of mathematical calculation and physical observations" (from "Science and the Modern World", A.N. Whitehead). According to the mainstream contemporary view, mathematics has its function in physics, which is identified with mathematical physics. Experimental physics has the functions of testing the physical theories and obtaining the physical properties.

In reality the relation between mathematics and physics is not that simplistic and obvious. A conflict between them arises due to the abstract nature of mathematics and the concrete nature of physics. "The point of mathematics is that in it we have always got rid of the particular instances, and even of any particular sorts of entities" (from "Science and the Modern World", A.N. Whitehead). On the contrary, the goal of physics is to establish the relations among particular physical entities and to predict the results for these entities using mathematical methods.

Awareness of some differences among physical and mathematical equations exists. "Physical equations are relations between physical quantities which, in general, state the balance of certain conceptual quantities as forces, energies, voltage drops, currents, momenta and so on" (from "Handbook of Engineering Fundamentals", edited by Ovid W. Eshbach and Mott Souders). Although this statement includes the relations among "conceptual quantities", in reality the only rule applied to the physical function distinguishing it from the pure mathematical equation is the rule of dimensional homogeneity, the rule of balancing the dimensions. The real difference among the mathematical and physical functions is much greater.

In mathematics no true or false question is asked in the initial description. The main purpose in physics is to construct functions that give verifiable results. That physical function has to establish the mathematical relations among the physical concepts as the entities. This approach creates a fundamental methodological and practical difference. We have to pay attention to the proper selection of physical concepts and establish reliable relations among them.

For the construction of a verifiable physical function we approach a structure as a geometrical whole. For, according to Max Planck, "it is impossible to obtain an adequate version of the laws for which we are looking unless the physical system is regarded as whole." The equations of the theory of elasticity that refer to a structure as a whole are the equations of the elastic deformations and of displacements. Let us examine these equations.

A physical function establishes the relations among the physical concepts as the entities. From this point of view, the linear equation of elongation, $e = NL/EA$, has variables and a constant that are physical concepts: the elastic force N, the length of the bar L, the cross-sectional area A and the modulus of elasticity E of the material. The same can be said of the similar equation of the angular deformation, $\theta = ML/EI_x$, where M is the mean value of the distributed elastic moment, L is the length of the beam, and I is the moment of inertia of cross-section.

On the other hand, the equation for the vertical displacement of a beam, such as $Y_{\max} = PL^3/48EI_x$, from this point of view is not a physical equation. Thus the variable and coefficient $L^3/48$ in the equation is not a single physical concept but an artificial mathematical construction, which combines parts of concepts of different origin. It is the part $L/8$ of the elastic moment $M = PL/8$, the part of geometrical description of the beam L, the part $L/6$ of the geometrical description of the resulting deformation $\theta = 6Y_{\max}/L$. This artificial equation is made with disregard for the physical meaning of the physical concepts. The relations are distorted in this equation. This makes it impossible to arrive at a meaningful inference and test the equation.

The same analysis can be applied to the equations of displacements in the popular computer-aided methods of finite-element analysis and the energy methods. The equations have so-called stiffness or flexibility matrixes. A matrix combines elements of different physical concepts, i.e., parts of the elastic forces and moments, geometrical descriptions of the structure, the modulus elasticity of the material, and parts belonging to the resultant deformation. Physical concepts can be presented in an equation in different ways, including a presentation in the form of matrix. But no matter how a variable is presented in the physical equation it should have physical meaning and participate in the relation as a whole concept. In that sense a physical equation is a more rigid construction than a pure mathematical equation. But what practical difference does it makes if it is a mathematical or physical equation if the end result of the calculations is the same? The physical equation must establish valid relations among physical concepts, which is impossible if the physical content of the variables is not taken into consideration.

In mathematical logic, "All you assert is, that reason insists on the admission that, if any entities whatever have any relations which satisfy such-and-such purely abstract conditions, then they must have other relations which satisfy other purely abstract conditions" ("Science and the Modern World", A.N. Whitehead). Such logical structure applies to mathematics unconcerned with establishing as true or false the physical assumption or the conclusion. From the point of view of formal logic, a hypothetical initial equation is correct if its derivative is also proven to be correct. The function is not necessarily correct just because its initial description is supported with the facts. The derivative equation, in this view, also should be supported by the facts. From the point of view of physics, the system of the equations at two levels, basic and derivative needs to be proven correct.

Such an approach suggests that even a function which has as variables single physical concepts may have invalid inferences. For example, the equation of elongation $e = NL/EA$ has the derivative $\mathrm{d}e/\mathrm{d}A = -NL/EA^2$. This derivative in the interval of the limit of elasticity of the material gives a calculated rate value that is inconsistent with our physical views and experiences, i.e., in the interval of the limit we should witness a rapid increase in the rate of change of deformation. Instead, the equation may give values of rate that do not indicate drastic changes. The equation of elongation in its present form is a linear equation that cannot predict the limit. Such an equation is not an adequate description of the elastic behavior. It is our job to select variable concepts in the equation that allow us to correctly analyze the changes of deformation.

The same analysis can be applied to the equation of angular deformation of a beam, $\theta = ML/EI_x$. The derivative equation, which should describe the rate of change of deformation due to the change of the moment of inertia I_x, that is, $\mathrm{d}\theta/\mathrm{d}I_x = -ML/EI_x^2$, does not predict actual changes in the interval of elastic failure. The relations as they are described in the basic equation do not exist in the interval of rapid changes.

For us to predict the changes, we must choose an independent variable that allows for the equation of deformation to exist in the domain of the tangent function, which serves as the derivative of the function of deformation. The properties of the tangent function are known and that lets us calculate the elastic behavior mathematically. On the other hand, the derivative of a hypothetical function is also hypothetical. In such a case we have no means to verify a theory. In physics we shall consider a logical structure that can be verified.

All physical concepts in the equation of deformation are entities representing main categories of the phenomenon, i.e., elastic force or moment, geometrical stiffness and modulus of elasticity. For example, in the case of deformation of a simple beam, we will consider the equation of angular deformation $\theta = M/ER$, where $R = KI_x/L$ is a function here, but is a single characteristic in the equation of deformation. The coefficient K represents the effect of the specific features of the beam design on geometrical stiffness. The coefficient can be obtained experimentally. In simple cases $K=1$ and the geometrical stiffness is proportional to the moment of inertia of the cross-section and inverse to the length of the beam. Geometrical stiffness is a comprehensive physical characteristic of a structure with respect to the phenomenon of elastic deformation.

The same physical change of cross-section with the different presentations gives different inferences. We have to choose the adequate description that can be verified. We can select the complex variable R in such a way that the function exists in the interval of rapid changes of deformation. In this interval the prediction can be verified or proven false.

It is equally important to treat each of the variables in a physical equation as a physical entity belonging to its physical category. For example, the Infinitesimal Theory of Elasticity has the equation of stress–strain equivalence, $\sigma = E\varepsilon$, Hooke's Law. This equation carries important conceptual meaning and is an important experimental tool. In the tensile test of a material we can, by measuring the stress and corresponding strain, calculate the modulus of elasticity of the material E. Also, if the modulus of the material is known, then by measuring deformation we can calculate the corresponding stress. The equation tells no more. But, the concepts of this equation have been extended in order to describe the elastic relations in a whole structure. Stress is presented as $\sigma = N/A$, strain as $\varepsilon = e/L$. The equation of equivalence after substitution became the equation of elongation, $e = NL/EA$. Mathematics allows such a transformation. But from the physical point of view two assumptions are made that should be viewed with a critical eye. First, the law for an infinitesimal unit was extended on the whole structure. In reality the geometry of a structure plays an important role that affects the relations. Secondly, after substitution of concepts into the equation the relations among variables were changed. The validity of the new equation should be tested mathematically and experimentally. The concepts in this equation should be selected in such a way that they aid in getting to the next logical step, the derivative equation which also should be proven correct.

8.1 Summary

In physics the rules of pure mathematics undergo some changes because the mathematical forms represent physical substance. This changes the degree of freedom for the mathematical transformations. It is important to select the variables in a physical equation with concern for their physical meanings.

The selection of an independent variable is especially important. The independent variable concept has to relate to the function. The dimensional level of an independent variable should be the same as the property to be analyzed. In our case, it is elastic deformation that dictates the dimensional level of geometrical stiffness.

A mathematical function is clearly distinct from a physical function. In mathematics we operate with abstract symbols. In physics we establish functional relations between concrete physical concepts, which are present in the equation as entities. A physical function is concrete, unique and does not allow the substitution of the variables.

The variables are selected to satisfy the two-level mathematical–logical structure of basic and derivative equations. Such structure allows for verification of a physical function. A physical function needs to be tested mathematically and experimentally.

The logical test consists of creating a two-level system of equations for which the derivative function has a character that is consistent with changes in a phenomenon. The corresponding basic equation, on the other hand, can be tested experimentally. This method creates the proven system for a physical function.

9. On the nature of the limit of elasticity

Physical theory deals not only with the construction of physical functions, but also with the establishment of the domain of application for these functions. The property that is associated with the domain in the theory of elasticity is the limit of elasticity. To predict the limit it is important to understand its origin. In the foundation of this process lies the general point of view on relations between the whole and its parts.

Classical mechanics views the whole as a sum of its infinitesimal parts. The literal division of a matter brings us ultimately to its atomic–molecular structures. Fundamental properties of a whole are identified with the properties of the infinitesimal parts of a material. “Theoretically, the strength of a material should be reflected by the forces at the atomic level. However, because of defects in the structure, the practical strength of materials is several orders of magnitude less than theory would predict” (from “Engineering Design”, Joseph H. Faupel and Franklin E. Fisher).

Observations and experiments do not support this point of view. Structures made of the same material in general have different elastic limits. Currently destructive testing is the common method for finding the limit of a particular structure. The limit of elasticity is known to be several orders of magnitude less than theory would predict based on the forces acting at the atomic levels. Also, “fundamental data obtained in a test on material are affected by the method of testing and the size and shape of specimen” (“Handbook of Engineering Fundamentals”, edited by Ovid W. Eshbach and Mott Souders).

Here is a different point of view on the origin of the limit of elasticity of a structure. We may consider a whole as consisting of autonomous subsystems. Each subsystem has its own properties depending on the geometry of the subsystem. Of practical importance is the division of a body on three levels: it atomic–molecular structure, the macrostructure of a material, and the body itself as the geometrical whole. The forces acting at the atomic–molecular level are too strong to be destroyed by the common elastic forces. But the forces at the level of the macrostructure of material and the limit generated by the geometry of a structure are of comparable values. The limit of elasticity, which manifests itself in the destructive test commonly performed for the structure limit, is of a relative nature; it can be the limit of the material or the limit generated by the geometry of the structure, whichever is lower. Both limits should be known in order to predict the common limit. The methods for obtaining these limits are different. The limit of a material can be found experimentally in a way similar to the current tests on the special specimen. The limit generated by the geometry of a structure can be found mathematically with the derivative of the new equation of deformation and the coefficient of elastic stability.

This theory has been tested, though on a small quantity of specimens. Tensile tests of specimens of different lengths cut off the same rod, $d = 0.5$ in, showed that these specimens had different limits. The limits were predicted correctly with the coefficient of elastic stability, $C_s = 3.7$ (tan 75°). The specimen with a calculated limit exceeding the limit of the material was destroyed when the tensile force reached the limit of the material. The stress–strain diagram (Illustration 4, below) illustrates this test. Thus, the relative character of the limit of elasticity was accepted as a part of the new theory of elasticity. The method of optimization of structures was devised based on this new theory.

One's point of view on the relation of the whole to its parts is important when building a theory. Thus, the Infinitesimal Theory of Elasticity is focused on the infinitesimal unit of a structure rather than on the structure as a whole. The assumption is made that the whole is a simple sum of its parts. The fundamental concepts of this theory are second-order tensors. The deformation is presented with the strain tensor. The elastic force is presented with the stress tensor. The limit ascribed to the material points to Hooke's Law, $\sigma = E\varepsilon$.

> Hooke's Law. It has been found by experiment that a body acted on by external forces will deform in proportion to the stress developed as long as the unit stress does not exceed a certain value, which varies for different materials. This value is the proportional limit.
>
> *(Handbook of Engineering Fundamentals, Ovid W. Eshbach and Mott Souders)*

Although this concept is presented as a fact or a law in reality, it is a quite selective approach to the facts. A tensile test of identical standard specimens made of different materials shows that they have different limits. The limit depends on the material. Different materials have different limits. A tensile test of specimens having different dimensions (lengths and cross-sections) but made of the same material shows that the specimens also have different limits. The limit depends on the geometry (size and shape) of a structure. These facts are known but the current point of view on the limit of elasticity as a property of a material prevents a scientific solution.

> Fundamental data obtained in a test on material are affected by the method of testing and the size and shape of specimen. To eliminate variations in results due to these causes standards have been adapted by ASTM, ASME and various associations and manufacturers
>
> *(Handbook of Engineering Fundamentals, Ovid W. Eshbach and Mott Souders)*

It is impossible to eliminate the differences in size, shape and method of loading for the infinite number of structures. The theory should be built in acceptance of the existence of the individual limit of a structure.

9.1 Summary

All systems dynamic and static are governed by a force that is characteristic for the system. For a static system this is an elastic force that keeps a structure as a whole. The elastic force reveals itself when external forces applied to a structure cause deformation of the structure. The value of the limiting elastic force, which does not lead to a permanent change of a structure, depends on the geometry of the structure and the elasticity of the material. The structure-specific limit should be known because it is usually the limit for the actual stresses acting on a structure. The limit of elasticity of the material comes to the fore in cases where the geometry of a structure allows higher stress than the material of the structure can withstand. Both limits should be known for the purpose of making a reliable design.

10. The stress–strain diagram

The diagram of the stress–strain relationship shown in Illustration 4 illustrates the results of a tensile test that was conducted on four specimens cut off the same rod, so they were of the same material and had identical cross-sections. The specimens were distinguished only by their lengths.

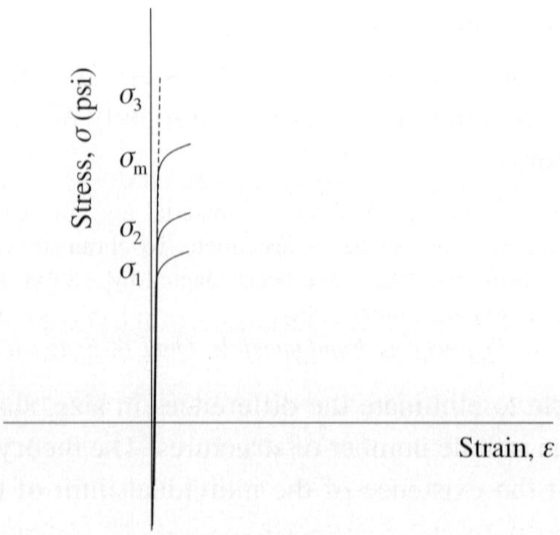

Illustration 4.

The actual elastic limits for specimens #1 and #2, i.e. σ_1 and σ_2, coincided with the calculated individual elastic limit. For specimen #3, failure occurred when the stress reached the limit of the material, σ_m that was less than the calculated individual limit σ_3. In this case the material presented the limitation.

11. On the nature of proof in physical theory

A satisfactory physical theory allows us to make predictions of physical facts by using mathematical descriptions of the relations. A physical theory is dual in nature: physical and mathematical. The physical part refers to the physical concepts. The mathematical part refers to the description of the relations among the concepts. Verification of a theory is also necessarily two-fold, empirical and mathematical. The necessity of empirical verification is recognized. Even more so, verification of a physical theory is identified with empirical confirmation of the predictions. "Verification is a process of confirmation of an hypothesis by showing that its predictions agree with the facts" ("College Logic", Alburey Castell).

In reality, confirmation of a physical theory cannot be done only through empirical verification. This impossibility is inherent in the dual nature of physical theory. We deal with two kinds of physical equations: the basic equation that operates only with the absolute values of the physical concepts, and the derivative equation that shows relations between the function and an independent variable and has a relative value as its result. We can measure and verify the absolute values of the concepts, which are particular physical characteristics. But it is impossible to measure the relations among them. A relative value includes the assumption that we selected the variables correctly and that the particular relation exists as described. There is no means to prove that empirically. This problem can be solved if the derivative is identified with a function whose properties are known, such as the tangent function. Then, we have no need to verify the derivative function itself. But it will place demands on the selection of the variables, and especially the independent variable, in order to satisfy both basic and derivative equations.

The independent variable can be selected mathematically from the logical structure of a derivative function such as the tangent function. The tangent has two intervals, the interval of slow changes and the interval of rapid changes. An independent variable should be chosen so that the basic equation exists in both intervals. A test by "significant figures" has been devised to assist in and verify the selection of the independent variable, $1 < \mathrm{d}Y/\mathrm{d}X < 10$. The limit for the physical function can be found in the interval of significant figures for its derivative. The domain of elastic behavior corresponds to the interval $0 < \tan \alpha < 4$. This interval may be called the "domain of physical existence" because beyond this interval the magnitude of the tangent function increases rapidly towards infinity, which is incompatible with the physical existence of stable relations or relations at all.

This method of testing the relations within a domain of the tangent function applied to the theory of elasticity introduces the important characteristic defined as "geometrical stiffness". Selected as an independent variable it has roots in the phenomenon of elasticity, because it is the real observable compound characteristic, and at the same time it satisfies the rigid logical–mathematical structure of a theory.

11.1 Summary

A two-level logical–mathematical structure establishes a connection between theory and phenomenon. This system can be verified or proven false. Proof of a theory is a two-fold process. The basic equation can be verified experimentally. The rate of change of a function is a mathematical property of that function; therefore the complete verification of a system of equations can be achieved only by adding mathematical confirmation of the system. To that end the a derivative function is selected which already has mathematically known properties. The derivative function also should adequately describe the character of the physical changes of the phenomenon. If for every point of a continuous basic function the derivative values belong to a function that is known to be a correct function then such structure can provide proof in the full interval of the function.

12. History of the theory of elasticity

The history of the ideas forming a theory is important for understanding these ideas, and for new ideas coming into the theory and changing its direction. In this work philosophy and the methods of the theory of elasticity undergo fundamental changes and therefore it is important to see the new principles in a historical perspective.

The physical point of view in classical mechanics is that a whole can be presented as a sum of its infinitesimal parts. The main limiting characteristic of the whole, the limit of elasticity, is an attribute of the part. In this work we consider a whole as it is built at the different levels of the structural division: the atomic–molecular level, the level of the macrostructure of a material, and the body as a geometrical whole that is also part of this division. The limiting characteristic of the whole is obtained by comparing characteristics at these levels.

The point of view on mathematics in physics has been reviewed in this work. A common point of view is that physical results can be obtained upon application of general mathematical rules to the function. This work shows clearly that the physical result depends on the physical content of the variables. Therefore, some general rules of pure mathematics cannot be applied to the physical function. The rule that the function of a function can be viewed as a single function is not applicable to the physical function, for the transformation changes the physical relations among the variable concepts in the function.

The logical structure and the objective validation of the new theory come with some retreat of the main argument of classical logic "If proposition P is correct, then conclusion Q is also necessarily correct." The point of view in this work is that Proposition P can be proven true only if it corresponds to the an independently known universal function that can serve as an inference Q, and only in the domain of possible existence of the physical phenomenon.

Finally, elastic relations have a limit, and therefore current linear theory, which cannot by its nature describe a limit, was replaced by the Non-Linear Theory of Elasticity. The physical idea that deformations are small in comparison with the dimensions of a structure, so that their relations can be described with a linear function, may lead to the erroneous physical idea that the limit of elasticity is not part of the described relationship. A physical point of view distinguishing the new theory is that deformations may have small values but the differences among them are clear at their own dimensional level. An analysis of change of deformation is necessary to establish the limit and select the optimal structure.

The science we are dealing with considers the principles of strength of materials and the theory of elasticity. The first formulation of its problems is associated with Galileo, who gave a general direction to this science with his seventeen propositions. The first of them is "Among an infinite number of homogeneous and similar beams is only one of which the weight is exactly in equilibrium with the resistance of the base of fracture. All others if of a greater length will break; if of a less length will have a superfluous resistance in their base of fracture." Second proposition: "The problem is to find the form of the generating curves so that the resistance of a section may be exactly equal to the tendency to rupture at that place (1638)" (from "A History of the Theory of Elasticity and of the Strength of Materials", Isaac Todhunter and Karl Pearson, 1884). Galileo assumed that each beam has its own individual strength, or resistance to fracture. This resistance, in Galileo's view, depends on the material. He considered homogeneous beams. Resistance depends on the load/weight and the length. Generating a mathematical relation/curve among those factors, the cross-section of a beam in equilibrium with resistance can be found. Galileo assumed that among the structures made of the same material there is only one structure that has optimal correlation between length and cross-section. Galileo treated solids as inelastic and did not generate the curves of resistance. The problem of finding the cross-section of a beam is known as Galileo's Problem.

The idea of an individual limit that can be found mathematically unfortunately has been abandoned in the modern theory of elasticity. The limit for a structure is identified with the constant characteristic of a material, i.e., the limit of elasticity of a material. The common technique for finding an individual limit of a structure until now has been the destructive test. The Non-Linear Theory of Elasticity that is presented here has a criterion for the analytical determination of the individual limit.

Robert Hooke laid down the physical foundation for the theory of elasticity. In his work "De Potentia Restitutiva" (London, 1678), he stated that eighteen years prior he had first found out the theory of springing bodies. "That is, the power of any spring is in the same proportion with the tension thereof. From all which it is very evident that the Rule or Law of Nature in every springing body is, that the force or power thereof to restore itself to its natural position is always proportionate to the distance or space it is removed therefrom. Respect being had to the particular figures of the bodies bended, and the advantages or disadvantages ways of bending them" (from "A History of the Theory of Elasticity and of the Strength of Materials", I. Todhunter and K. Pearson).

From this extract, Hooke's point of view is that all bodies, of all materials, have elastic properties. The power of resistance is proportional to the deformation, with respect to the geometry and the boundary condition. "Hooke did not make any application of it to the consideration of Galileo's Problem. Hooke's Law provided the necessary experimental foundation for the theory." (from "A Treatise of the Mathematical Theory of Elasticity", by A.E.H. Love, 1892). Hooke did not define the domain for his law.

The modern interpretation of Hooke's Law is as follows: "It has been found by experiment that a body acted on by external forces will deform in proportion to the stress developed as long as the unit stress does not exceed a certain value, which varies for different materials" (from "Engineering Fundamentals", edited by Ovid W. Eshbach and Mott Souders). Thus, the domain of the law of elasticity for a body in this interpretation is a constant of the material.

James Bernoulli made the first investigation of the elastic line in 1705. Leonhard Euler assumed a result in his investigation of a column. He made the assumption that failure of a column is due to sidewise bending. This assumption is not generally true even for a column. However, the feature of his hypothesis concerning the failure criteria is important. Strength of a column, or the critical load, is not a constant of a material. It is, rather, a function of geometry, modulus of elasticity of material and the boundary condition. The strength of a column is an individual value that depends on the design itself.

"In the course of the nineteenth century, mechanistic theories of a different kind were constructed. In these, matter was treated as though it were continuous. The reason for the retreat from the corpuscular theory of matter was that this theory required complicated mathematics and did not provide reliable predictions. On the other hand, the treatment of continuous deformable media requires the application of a partial differential equation. After the development of this kind of equation by the mathematicians of the 19th century Cauchy, Poisson, Jacoby, it became possible to study mathematically the statics and dynamics of continuous deformable media, and the mathematical theory of elasticity was founded." (Rephrased from "The Rise of a New Physics" by A. De'Abro).

The formulation of the general equations of the linear theory of elasticity was made by Navier in 1821. It is considered the foundation of the modern theory of elasticity. Philosophical and mathematical approaches in linear theory are that the whole can be presented as a sum of its infinitesimal parts with linear equations. At the end of the 19th century it was clear that the linear theory of elasticity could not solve the majority of practical problems.

Traditional strength theories consider the domain of Hooke's Law outside the law as a material's property. All existing strength theories are based on the assumption that the strength of a structural member and, therefore, the limits for design are constants of the material. The limiting maximum stress, limiting maximum strain, or limiting strain energy do not depend on the load, geometry of design, modulus of elasticity and boundary condition. The limits for design calculations are the characteristics of the material rather than the characteristics of the particular design.

In "Engineering Fundamentals" edited by Ovid W. Eshbach and Mott Souders, the theories of the domain of elastic relations or in other words the theories of limit are described as follows.

> The determination of the magnitudes and directions of the principal stresses and strains is carried out for the purpose of establishing criteria of failure within the material under the anticipated loading conditions.
>
> *Maximum-stress Theory* (Rankine's Theory). This theory is based on the assumption that failure will occur when the maximum value of the greatest principal stress reaches the value of the maximum stress σ_{max} at failure in the case of simple axial loading.
>
> *Maximum-strain theory* (Saint Venant). This theory is based on the assumption that failure will occur when the maximum value of greatest principal strain reaches the value of the maximum strain ε_{max} at failure in the case of simple axial, loading.
>
> *Strain-energy Theory*. This theory is based on the assumption that failure will occur when the total strain energy of deformation per unit volume in the case of combined stress is equal to the strain energy per unit volume at failure in simple tension.

The linear theory of elasticity considers an individual structure. The equation of the elastic line refers to the change of shape of a structure (physical body). The non-linear theory of elasticity considers a family of similar structures and the change of behavior of a structure depending on its position in the set. The equation of deformation in the non-linear theory of elasticity refers to the mathematical curve describing the relations in the set rather than to the physical curve of a body.

For constructing and proving the non-linear theory of elasticity it was necessary to apply modified rules of logic. These rules explain the connection between the cause, i.e. the work of the external force, and the effect, i.e. the phenomenon of elasticity. We explained the necessity of a two-level mathematical structure for completely describing both the relations in the deformed body and the applicable relations in a set of similar structures. The selection of variables has been explained as a logical necessity to satisfy a strict logical structure. The main purpose of logic is both justification of mathematical structure and justification of the selection of terms of a theory. Definitive logic satisfies the demand of a physical theory in its logical consistency. The empirical verification of such consistent theory proves its existential validity.

While the Non-Linear Theory of Elasticity is different from the traditional theory, it should be considered as a continuation of science, whose physical concepts, mathematical descriptions and methods constitute the body of the linear theory of elasticity. The roots of the new non-linear theory are in the existing knowledge, though the changes are significant.

13. On the principles of the theory of elasticity

Ordinarily we think about scientific theory as based on organized objective knowledge of a phenomenon drawn from the facts. In reality physical theory is knowledge acquired and organized from a certain philosophical position. In modern physics this position is defined as atomism. The essence of atomism lies in its method of studying the properties of a phenomenon as a whole by dividing the whole into the constituting elemental blocks of matter and studying the properties of these blocks. The properties of a whole commonly are identified with the properties of the elemental block such as the atomic–molecular structure.

Atomism in physics was originated in the 17th century and it has proven to be a valuable method for studying the properties of infinitesimal parts. The Infinitesimal Theory of Elasticity is based on that assumption. However, this theory is not successful in predicting the most important characteristic of a structure as a whole, i.e. the limit of elasticity of a structure.

However, atomism is not the only possible philosophical platform for science. The different approach here considers a more complex division of the whole into structurally identifiable parts at the different levels of the division. For example, an elastic body can be divided at the levels of the atomic–molecular structure, the macrostructure of a material, and the body as a geometrical whole being also a part of this division. Then the whole is more than the sum of its parts.

The scientific process of acquiring knowledge consists not only of analysis but of synthesis as well. Synthesis is conceptually different from analysis. In synthesis we consider not the material parts of the division but, rather, the properties of these parts. The property of the whole is usually not the property of the smallest blocks of the division. It is not necessarily the property of the physical whole. The property of the phenomenon as a whole can be established by comparing the corresponding properties of the parts. The limit of elasticity of a structure is a relative characteristic.

For constructing and proving a physical theory it is necessary to consider the logical connection of the theory to the phenomenon. Modern physics maintains an agnostic attitude to the possibility of establishing such a connection. Here is a different point of view. The logical connection of a theory to a physical phenomenon can be established. However, we should not seek logic in nature, for logic is our means of connecting things. A logical structure that can be proven should be purposefully incorporated into a physical theory as its framework. The content of such logical structure is physical concepts. Physical concepts are selected to satisfy this strict structure. It is a way to obtain reliable physical results.

The non-linear theory of elasticity has a two-level structure of logical–mathematical connections between the basic equation describing deformation and the derivative equation describing the rate of change of deformations. The analysis of the content of a theory shows its dual origin, physical and mathematical. A theory has abstract concepts of the real physical properties of a phenomenon, which can be measured and verified. The derivative equation has the relative value of change of deformation as its result. This value, in general, cannot be measured but can only be calculated. After calculation the rate corresponding to the elastic failure can be confirmed experimentally. The logical system can be proved if the derivative of the basic equation corresponds to a known function in the full physical domain of a phenomenon.

Proving a theory is also a two-fold process. The necessity of empirical verification is recognized. Even more, confirmation of a physical theory is identified with the empirical verification of the predictions. In reality, confirmation of a physical theory cannot be accomplished by using empirical verification only. This impossibility is inherent in the dual nature of physical theory. Absolute values of concepts can be measured and verified experimentally. But it is impossible to measure the relations between the concepts. The relative value in the derivative equation includes the assumption that we selected the variables correctly and that the particular relation exists as it is described. Only employment of the logical structure of a theory for verification can validate a theory. This problem can be solved if mathematical inference from the basic function is identified with a function describing the pattern of relation whose properties are known, such as the tangent function. Then the empirical proof would not be needed for the inference as classical logic suggests, but rather for verification of the basic equation that is possible.

The new methodology also requires a different approach to the application of the rules of mathematics in a physical equation. Mathematics can be a useful tool in physics when applied purposefully and with respect to the physical content of the variables. Otherwise, mathematics can be a source of fundamental errors in physics. A physical function is clearly distinct from a pure mathematical function. A mathematical function operates with abstract variables, while a physical function establishes the relations among particular physical concepts, which are treated as entities. This position restricts the rules of mathematics in physics significantly. It is important to select the variables in a physical equation with concern for their physical meaning and with concern for the logical structure of a theory. The mathematical forms in a physical function have physical substance that changes the degree of freedom for mathematical transformations. Indiscriminate use of mathematical rules in physics may lead to distortion of the relations among the variables and to erroneous inferences.

Another significant distinction between the non-linear theory of elasticity and the linear theory is that the linear theory gives the mathematical description of an elastic line representing an individual physical body. The non-linear equation of deformation, on the other hand, is a mathematical curve representing the relations in a family of similar structures. Seemingly one description of elastic relations serves two different purposes. In essence the equation of deformation appears to be two quite different descriptions.

The point of view on the cause–effect relations is also of consequence for building a theory. The point of view in this work is that cause (external forces) belongs to one world and effect (the phenomenon of elasticity) is in another world and belongs to a structure. No logical deductive connection can be established between cause and effect. The logical deductive connections described in the theory exist only among the concepts pertaining to a body. These are relations such as that among deformation, elastic force, geometric properties of a body, and material elastic property. These concepts–characteristics are intimately connected. A change of one concept leads to the change of others. The system in which effect takes place is viewed as an autonomous system. All knowledge about a phenomenon can be found within the system itself. The balance that exists between external forces and elastic force is also important though it is not a functional relation but a logical inductive relation between cause and effect. The external force performs a work that is equivalent to the elastic potential of a body. From this point of view Hooke's law, $\sigma = E\varepsilon$, is a symbolic description demonstrating equivalence of two properties of an elastic field at the infinitesimal level.

Elastic relations have a limit. A function that can describe such relations is a non-linear function. Also, only non-linear function may satisfy a logical need in the basic and derivative equations for proving the validity of a theory.

These principles have been applied to building and proving the Non-Linear Theory of Elasticity and the new patented Method of Optimization of Structures.

13.1 Summary

Physical theory is dual in nature: physical and mathematical. It refers to all features of a theory, namely, its mathematical description, its logical structure, its physical concepts, methods of verification of the theory, and a method of establishing the domain of the theory.

The logical–mathematical structure of a theory has two types of equations, a basic equation connecting physical concepts and a derivative equation establishing the relation between function and independent variable.

Physical concepts are the abstract representations of the physical properties of a phenomenon. They are selected to satisfy the rigid logical–mathematical structure.

Verification of the theory is mathematical because the rate of change of deformation is a mathematical property of the function of deformation. But the basic function can be confirmed experimentally.

Physical theory exists in some domain. Establishing the limits is also a two-fold process. On the one hand, the derivative equation logically shows the limit in the interval of rapid changes of the tangent function. On the other hand, matter/material presents its limitation. Comparative analysis of these limits shows the limit for a concrete structure.

Engineers deal with theories which are similar in their essence. A change is proportional to the force acting in the system and inversely proportional to the resistance of a system to this particular change. The descriptions of such relations also can be similar for the various physical phenomena.

United States Patent 5,654,900 (August 5, 1997) Method of and Apparatus for Optimization of Structures

L.W. Ratner

Abstract. The invention is of a basic nature; the entire technical disclosure is new in the art of design. The new method is based on a new concept of strength and elastic stability. Classical mechanics views a body to be formed of small units of material. Such fundamental characteristic of a body as an elastic limit is identified with limit of material unit. It is scientific conclusion supported with facts that every structure has its own limit, which, in general, is different from the limit of material. New basic equation of elastic deformation establishes deformation – elastic force –geometrical stiffness relations. Geometrical stiffness characterizes the resistance to deformation based on geometry of the whole body. New method suggests a different approach to using mathematical equations and procedures. In order to obtain valid and verifiable inferences from an equation attention should be paid to physical meaning of the variables. The new equation of elastic stability establishes relation between elastic force and geometrical stiffness in the interval of rapid change in deformation. The invention presents the new non-destructive method of design. Major problems in design process were formulated in 17th century, and yet have not been solved. And, although, the problems are the same, their economical and social impact in technological society of the 20th century is quite different.

1. Background of the invention

1.1 Field of the Invention

The present invention relates to the art of designing load-bearing design members and, more particularly, to the method of using data of deformation and geometrical stiffness for calculating the elastic limit and corresponding optimal dimensions of a member.

The present invention also relates to non-destructive testing resistance of a member to elastic deformation and, more particularly, to the testing of geometrical stiffness of a member.

The invention also relates to the testing of mechanical properties of the structures and materials.

1.2 Description of the Prior Art

The prior art of design is based on well-known theories of strength such as maximum-stress theory, maximum-strain theory, and maximum strain-energy theory. A physical concept underlying these theories is that material limits the application of Hooke's Law of elasticity,

$$\sigma = E\varepsilon. \quad (1)$$

According to the most common maximum-stress theory member is considered to be reliable if maximum stress in the member is less than proportional limit of the material.

> Hooke's Law. It has been found by experiment that a body acted on by external forces will deform in proportion to the stress developed as long as the unit stress does not exceed a certain value, which varies for the different materials. This value is the proportional limit"
> *("Handbook of Engineering Fundamentals", 3d Ed., p. 489, Eshbach and Souders)*

Then, the art of calculating dimensions of a member follows the theory. Stresses in the member can be obtained analytically or by measurement. Test of material using the standard specimen gives mechanical properties of the material such as proportional limit, elastic limit, ultimate strength, and modulus of elasticity of material. Maximum stress in the member then compared with proportional limit of the material for calculating the cross-sectional characteristics or correcting them. For example, in case of tension,

$$\sigma = \frac{N}{A} \leqslant [\sigma_p], \qquad A \geqslant \frac{N}{[\sigma_p]}. \tag{2,3}$$

In the case of bending,

$$\sigma_{max} = \frac{M_{max}}{Z} \leqslant [\sigma_p], \qquad Z \geqslant \frac{M_{max}}{[\sigma_p]}. \tag{4,5}$$

Presently, there is no universality in the theories of strength. In contrary to the general strength theories the theory of buckling is based on assumption that critical buckling load or stress does not depend on the critical characteristics of the material, but depends on geometry and modulus of elasticity of material only.

> Equation 9.15 [$F_{cr} = \pi^2 EI/4L^2$] is known as *Euler's column formula* and indicates that the critical buckling load is not a function of the strength of the material (yield and ultimate strengths are not involved) but only of the elastic modulus and geometry. *("Engineering Design", by Faupel and Fisher, 2nd Ed., p. 568)*

The prior art of design has great achievements. However, it has major flaws as well. The main disadvantage of the prior art is that strength theories do not corroborate well with the physical evidence. Thus, the strength theories contradict to overwhelming evidence that critical for a structure load or stress depends on geometry of design and modulus elasticity of material and not a function of the material strength.

This is true for buckling and all general cases of deformation as well. Confirmation to this point of view can be found in the tests of materials and structures.

> Fundamental data obtained in a test on material are affected by the method of testing and the size and shape of the specimen. To eliminate variations in results due to these causes, standards have been adapted by ASTM, ASME and various associations and manufactures.
> *("Handbook of Engineering Fundamentals", 3d Ed., p. 566, Eshbach and Souders)*

It is obvious that eliminate differences in the size, shape, and method of loading is impossible in the structures other than specimen. Different structures made of the same material have different limits.

The standards and tests of the structures are of some but insufficient help. Common physical foundation and the equations describing relations between critical for the design load and geometry of the design must be developed.

The buckling empirical formulas developed for the different practical cases are not applicable for general cases of bending, tension, torsion. In fact, these formulas are not very reliable even for cases of buckling. The buckling formulas are developed with the assumption that failure of the columns, for example, occurs due to the sidewise bending.

> This assumption is not true for very short columns, nor is it true for columns of medium length such as usually needed in practice. There is no exact formula which gives the strength of a column of any length under an axial load.
> *("Handbook of Engineering Fundamentals", 3d Ed., p. 529, Eshbach and Souders)*

The theories of strength remained hypothetical for centuries. And design technique became more and more complicated due to uncertainty in the art of design.

> Here the continuing trend towards lighter and thinner structures associated with the use of high strength material is bringing problems of elastic stability increasingly to the fore. This has long been the case in the aerospace field, but it is now rapidly extending to ships and to high-rise buildings. And as designs become even more efficient the engineer will be faced with even more instabilities demanding the sophisticated treatments
> *(A General Theory of Elastic Stability, 1971, London, p. 48, J.M.T. Thompson and G.W. Hunt)*

The present invention in the art of design is based on a new and different concept of strength. According to this concept each structure has an individual proportional and elastic limits which, in general, are different from the limits of the material. The limit for a structure depends on the resistance of a structure to elastic deformation. Physical characteristic describing this resistance is called "stiffness". Stiffness depends on elasticity of material (E), geometry of design and boundary conditions.

A part of the stiffness, which is a function of size, shape, specific design features and boundary conditions, is singled out and described as a new important characteristic of a structure called "geometrical stiffness". Then, general equation of elastic deformation can be written as following.

Total elastic deformation is proportional to the force distributed in the structure and inversely proportional to the geometrical stiffness and modulus elasticity of material,

$$D = \frac{F}{ER}. \tag{6}$$

Thus, in case of tension,

$$e = \frac{N}{ER} \quad \text{where } R = \frac{KA}{L}. \tag{7,8}$$

In case of bending total angular deformation,

$$\theta = \frac{M}{ER} \tag{9}$$

where geometrical stiffness

$$R = \frac{KI}{L}. \tag{10}$$

Geometrical stiffness in the equations of elastic deformation is presented as a physical entity.

The equations of deformation in the prior art are different and the difference is not formal but of practical importance. Presently, in case of tension,

$$e = \frac{NL}{EA}. \tag{11}$$

In case of bending the general elastic line equation is $M = +EI * \mathrm{d}Y^2/\mathrm{d}x^2$. From the general equation the equations for the different specific cases are developed. For example, for the simple beam with concentrated load at the center,

$$Y_{\max} = \frac{PL^3}{48EI}, \qquad \theta = \frac{PL^2}{16EI}. \tag{12,13}$$

For the purpose of optimization of dimensions it is necessary to know how geometry, in particular size, affects deformation. The equation, which should show such effect, is a differential equation derived from the equation of elastic deformation. Scientific logic suggests that if an equation represents the relations among components correctly then the derivative differential equation will be also correct. However, it appears that differential equations derived from the existing equations of deformation are incorrect. For example, $d\theta/dI = -PL^2/16EI^2$ does not describe the rate of change of deformation depending on change of moment of inertia of cross-section correctly. The equations of deformation in the prior art are unsuitable for the purpose of optimization.

The new equations of deformation are different. The main components in the equation are the elastic forces distributed in the structure, the geometrical stiffness, and the total deformation. Though, each of these components can be presented as a function in equation of deformation said components presented as the physical entities. This holistic approach differs from the existing disintegrated approach when the equation of deformation became a mixture of elements belonging to the components of different physical origin. Thus, in the equation $Y_{max} = PL^3/48EI$ one of the L-s belongs to the bending moment, another to the geometrical characteristic, yet another to the resulting deformation. Such attitude of neglecting physical meaning of the components led to the flaws in representation of relations and in results. New equations describe the elastic relations more accurately. Differential equations derived from them are also correct. On example of a beam deformation–geometrical stiffness relation is presented graphically in the diagram θ vs. R (Figure 1). The diagram shows rapid increase of deformation in the interval proportional–elastic limit. Beyond this limit, an insignificant decrease in stiffness results in failure of elastic behavior. Likewise, increasing geometrical stiffness above the proportional limit does not improve elastic stability. The point R_a in the diagram shows the position of an actual geometrical stiffness of tested structure. The point R_o shows the position of an optimal geometrical stiffness for given force and material.

The rate of change of deformation due to geometrical stiffness is a reliable criterion for design optimization. The proportional and elastic

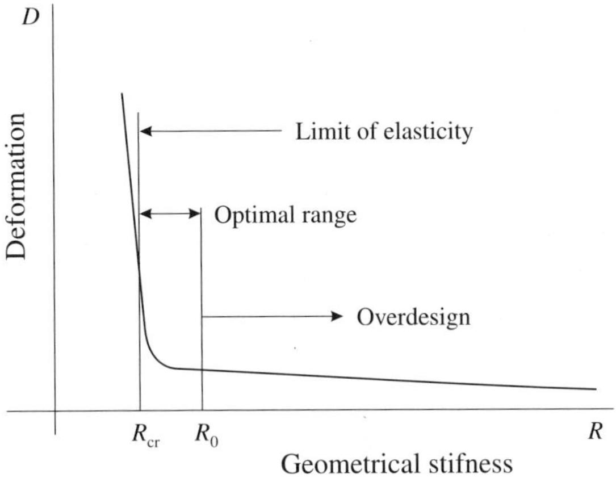

Fig. 1.

limits are characterized with the rate of change of deformation. In the interval of proportional–elastic limit the rate can be anticipated from $\tan\alpha = 1.0$ ($\alpha = 45°$) to $\tan\alpha = 3.7$ ($\alpha = 75°$).

The rate of change of deformation can be described with a differential equation derived from the equation of elastic deformation,

$$\frac{\mathrm{d}D}{\mathrm{d}R} = -\frac{F}{ER^2}. \tag{14}$$

The relations within limits are

$$\frac{F}{ER^2} = C_s, \tag{15}$$

where $C_s = \tan\alpha$ is the coefficient of elastic stability. Equation (15) is the foundation of elastic design. Thus, in case of bending,

$$\frac{\mathrm{d}\theta}{\mathrm{d}R} = -\frac{M}{ER^2} = -C_s. \tag{16}$$

Optimal geometrical stiffness is

$$R_o = \sqrt{\frac{M}{EC_s}}. \tag{17}$$

Geometrical stiffness of a beam is a function of moment of inertia of cross-section, length, specifics of a beam design and boundary conditions,

$R=KI/L$ (eq. 10) where K is a coefficient which counts effect of the specifics and boundary conditions on geometrical stiffness. The physical meaning of geometrical stiffness is clear from this description. The greater is the moment of inertia, the greater is geometrical stiffness. The greater is length, the less is geometrical stiffness. Different beams may have the same stiffness if they have the same ratio of moment of inertia to the length, $R=KI_1/L_1=KI_2/L_2$.

Absolutely different structures may have the same geometrical stiffness, $R=M/E\theta$. Considering geometrical stiffness as an entity, as a new property of a structure allows establish the standards of geometrical stiffness for the purpose of measurement. Note, that mode of deformation is considered in defining stiffness for bending, tension, torsion.

In order to correct geometrical stiffness of a beam, for example, one can change moment of inertia of cross-section. Here, $R_o/R_a=I_o/I_a$. The optimal moment of inertia is

$$I_o = \frac{I_a R_o}{R_a}. \qquad (18)$$

The problem of calculating the optimal moment of inertia with eq. (10) is in the fact that coefficient K in the equation, which accounts for the effect of specifics of design and boundary conditions, initially can be obtained only experimentally. Series of similar structures have common coefficient K.

In some cases the limit of elasticity of material may present the limitation for a structure. Therefore, maximum stress in the structure of optimal dimensions must be checked against stress allowable by the material,

$$\sigma_{max} = \frac{M_{max}}{S_o} \leqslant [\sigma_p]. \qquad (19)$$

The method of optimization of load bearing design members which is further described on the exemplary teachings is an essential general method. The method can be used for optimization of any type of design of any complexity and of any material. The method is directed to optimizing the series of similar structures by testing one representative. It is a very economical method. The new method will contribute to the safety and reliability of the structures and give savings on materials, labor, time.

The material presented makes clear the fundamental difference between the prior art of design and the new art, and the advantages of the new art. The prior art did not realize the existence of the individual limit of a structure. It makes the methods of the prior art deficient. Further, in order to choose proper dimensions it is necessary to know how geometry affects behavior of a structure. The rate of change of deformation is an indicator of elastic behavior. There is no knowledge of that in the prior art. Fixed criteria of limiting stress and limiting deformation in the prior art do not describe elastic behavior and they are unsuitable for the purpose of optimization.

There is no equation, which describes rate of change of deformation depending on geometry, in the prior art. The equation of elastic deformation that exists in the prior art cannot be used for that purpose for it does not describe relation deformation–geometry correctly.

A new property of a structure, i.e. geometrical stiffness, is introduced in the art of design in order to reflect the effect of geometry on elastic behavior correctly. The new equation of elastic deformation describes deformation–force–geometrical stiffness relations. The derivative equation describes the rate of change of deformation depending on geometrical stiffness. Both equations are essential for a scientific design process but are missing in the prior art.

An equation describing geometrical stiffness of a structure makes it possible to compare similar structures of different dimensions. It allows economical optimization series of similar structures after testing stiffness of a representative structure.

The new art challenges prior art. It makes it possible to compare structures, to predict behavior of structures, to make design process scientific rather than empirical.

Major differences between the prior art of design and new art are summarized in the Table of Comparative Analysis of Prior Art and the New Method.

Prior Art	New Method
1. Hooke's Law	
Strength of a structure is identified with the strength of material, $\sigma = E\varepsilon \leqslant [\sigma_p]$	Each structure has an individual elastic limit, which is different from the limit of material
2. Equations of elastic deformation	
Tension: $e = NL/EA$	$e = N/ER$, $R = KA/L$
Bending: $\theta = PL/K'EI$	$\theta = M/ER$, $R = KI/L$
Torsion: $\phi = TL/GJ$	$\phi = T/GR$, $R = KJ/L$
	In the equations of deformation, R is a single characteristic defined as geometrical stiffness.
3. Equations of elastic stability	
The equations that should describe how geometry affects rate of change of deformation are not developed	$N/ER_o^2 = C_s$, $M/ER_o^2 = C_s$, $T/GR_o^2 = C_s$, where C_s is the coefficient of elastic stability
4. Main criterion for design	
Certain characteristic of a material such as proportional limit is criterion for design	Certain rate of change of deformation is criterion for design. It is true within elastic limit of material
$\sigma = N/A \leqslant [\sigma_p]$, $A \geqslant N/[\sigma_p]$	$R_o = \sqrt{N/EC_s}$, $R_o = KA_o/L$
$\sigma = M/S_x \leqslant [\sigma_p]$, $S_x \geqslant M/[\sigma_p]$	$R_o = \sqrt{M/EC_s}$, $R_o = KI_o/L$

2. Summary of the invention

The invention is a new method of designing load-bearing structures. The method is based on a new theory of strength and elastic stability. According to this theory each structure has an individual limit of elasticity, which depends on geometrical stiffness of the structure. Geometrical stiffness corresponding to the limit provides elastic stability. The new method is a process of designing a structure of predetermined optimal geometrical stiffness. Optimal geometrical stiffness can be calculated with a new equation of elastic stability. This equation describes rate of change of deformation due to geometrical stiffness. The geometrical stiffness depends on the correlation between cross-sectional and longitudinal dimensions, specific features of a structure and boundary conditions. In most practical cases it is optimal cross-sectional characteristic, which should be found. The method contains a non-destructive test in order to find the coefficient K, which counts effect of specifics of a structure on geometrical stiffness. For similar structures the coefficient K is considered the same. The method allows optimization of dimensions of a series of similar structures by testing one representative structure.

The new method, which is claimed, comprises the steps of evaluating forces distributed in the structure and estimating arithmetic mean value of said forces; measuring total deformation caused by these forces; determining actual geometrical stiffness of tested structure using data of mean value of forces and total deformation, $R_a = F_m/ED$; calculating optimal geometrical stiffness required by mean value of the distributed forces with the new equation of elastic stability, $R_o = \sqrt{F_m/EC_s})$ where C_s is a coefficient of elastic stability; comparing optimal geometrical stiffness with actual geometrical stiffness in order to exclude coefficient K and to correct the cross-sectional dimensions so that optimal geometrical stiffness is attained, $R_o/R_a = I_o/I_a$. It is the purpose of the new method to obtain necessary stiffness. However, the structure of optimal dimensions must be checked against limit of elasticity of material in order to select the material properly.

The new method also directed to optimization series of similar structures. Compare optimal geometrical stiffness of the tested structure with optimal geometrical stiffness of similar structure using the equation describing stiffness one can calculate optimal dimensions of similar structure, $R_o/R'_o = (I_o/I_o^1) * (L'/L)$. The said method can be used with apparatus for measuring deformation in a structure, inputting the data in the computer for the processing the information with a new program as described and displaying results. The method can be used for optimizing beams, plates, shells, cylinders, columns, and shafts.

3. Description of illustrated exemplary teaching

Reference is now made to the drawings, which illustrate an exemplary teaching of said new method of and apparatus for optimization of

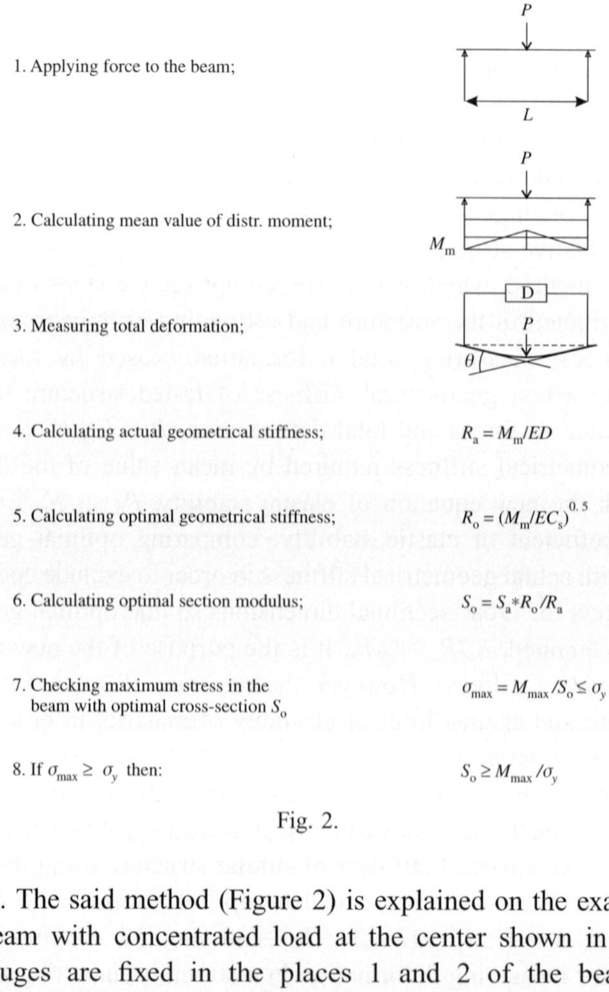

Fig. 2.

structures. The said method (Figure 2) is explained on the example of a simple beam with concentrated load at the center shown in Figure 3. Strain gauges are fixed in the places 1 and 2 of the beam 3 and connected to the apparatus 4. The apparatus has means for evaluating deformation, elastic forces, and geometrical stiffness. Said apparatus has also a calculating device with a program and display.

(1) Dynamometric (known) force P causes distribute along the beam bending moment shown in the moment diagram. Maximum moment, $M_{max} = PL/4$.
(2) Arithmetic mean value of distributed load, $M_m = M_{max}/2$.
(3) Total angular deformation, $\theta_a = \alpha_1 + \alpha_2$ is measured by means of a strain gauge device as a part of apparatus 4.

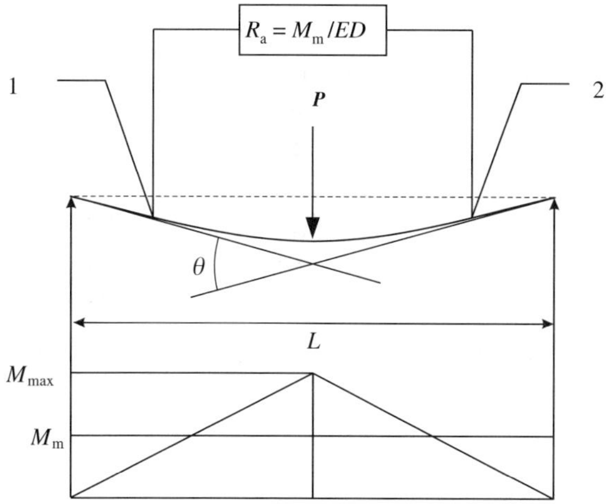

Fig. 3.

(4) Actual geometrical stiffness of the beam can be calculated using data obtained in steps 2 and 3, $R_a = M_m/E\theta_a$.
(5) Optimal geometrical stiffness can be calculated using an anticipated coefficient of elastic stability, $R_o = \sqrt{M_m/EC_s}$.
(6) Optimal moment of inertia of cross-section, $I_o = I_a * R_o/R_a$.
(7) Maximum stress in the beam of optimal dimensions must be less than allowable by the material stress, $\sigma_{max} = M_{max}/S_o \leqslant [\sigma_p]$.

In some more complex cases than this example the force distributed in the structure is difficult to calculate. Then, the following technic can be used:

(1) Applying to the structure an additional dynamometric force, P_d.
(2) Measuring total deformation from this dynamometric force, θ_d.
(3) Calculating arithmetic mean value of moment distributed in the structure, M_d.
(4) Calculating geometrical stiffness of the beam, $R_a = M_d/E\theta_d$. Note, that actual stiffness of tested/actual structure does not depend on the force applied to the structure. Greater force causes proportionally greater deformation.
(5) Measuring deformation caused by the actual external forces, θ_a.
(6) Calculating arithmetic mean value of forces or moments distributed in the structure, $M_m = E\theta_a R_a$.

(7) Calculating optimal geometrical stiffness, $R_o = \sqrt{M_m/EC_s}$
(8) Calculating optimal cross-sectional dimensions, $I_o = I_a * R_o/R_a$.
(9) Checking the maximum stress in the beam of optimal dimensions against elastic limit of material, $\sigma_{max} = M_{max}/S_o \leqslant [\sigma_p]$.

Although the test is simple it is unnecessary to repeat it in order to optimize similar structures. If the force in a similar structure is different then this force should be evaluated. The optimal geometrical stiffness of a the similar structure is calculated, $R_o = \sqrt{M'_m/EC_s}$. Optimal cross-sectional moment of inertia in case of beam, $I'_o = I_o(L'/L) * (R'_o/R_o)$, where R_o, I_o and L are the parameters of tested beam, and R'_o, I'_o and L' are the parameters of the similar beam.

Geometrical stiffness of a beam can be calculated using the equation of elastic deformation, $R_a = M_m/E\theta_a$. Or, geometrical stiffness can be measured by means of a resistant element such as a rheostat as a part of apparatus 4. The resistance of the rheostat is regulated accordingly to a value M_m/E representing a standard of known R_sQ_s. Measuring total deformation in the beam, its value is then compared with the value of standard resistance. The actual geometrical stiffness of a beam is $R_a = R_s\theta_s/\theta_a$.

The method described above on the example of simple beam can be successfully used for any type of design. The more complex a structure the more beneficial the method. Fixing strain gauges at some distance from each other at the structure and applying a known force we can measure total deformation and geometrical stiffness of this part of the structure as well. Force distributed in said part also can be measured by measuring strains. Then, optimal stiffness required by the actual force can be calculated. Geometrical stiffness of a structure can be corrected by changing one of the geometrical parameters. Usually this is cross-section. For example, for plates and shells the geometrical stiffness can be corrected by correcting thickness, $t_o^3 = t_a^3 R_o/R_a$. For columns the geometrical stiffness can be corrected by correcting cross-sectional area, $A_o = A_a R_o/R_a$.

Obviously, many modifications and changes may be made to the foregoing description of the methods and apparatus without departing from the core of the invention. The material presented makes clear the fundamental difference between the prior art of design and the new art embodied in this invention. The new art considers an individual limit of a structure unknown in the prior art. The individual limit is described with the coefficient of elastic stability unknown as it is in the prior art. The individual limit of a structure depends on geometrical stiffness of a structure, a property unknown in the prior art. Thus, the new methods of design are based on a new knowledge. The new methods are directed toward optimization of similar structures as well as optimization of an individual structure. There is no other method in the prior art which tries to or can achieve this purpose. This is because the criteria of fixed limiting stress and fixed limiting deformation do not describe behavior of structures and cannot be used for the purpose of optimization. The characteristic 'geometrical stiffness' is not developed in the prior art. It makes it impossible to compare the structures. Said new scientific methods allow us predict elastic failure of a structure and offers the most economical way of design the optimal structures.

Notation

Geometrical properties

A	area of cross-section
I	moment of inertia of cross-section
I_a	actual moment of inertia of cross-section
I_o	optimal moment of inertia of cross-section
I_p	polar moment of inertia
L	length
R	geometrical stiffness
R_a	actual geometrical stiffness
R_o	optimal geometrical stiffness
S	section modulus

Forces

F_m	arithmetic mean value of elastic force
M	bending moment
M_{max}	maximum bending moment

N	axial force
T	twisting moment
σ	unit stress
σ_{max}	maximum stress
σ_p	proportional limit
σ_y	elastic limit
τ	shear stress

Deformation

e	total deformation in tension or compression
ε	unit deformation/strain
θ	total angular deformation in bending
φ	angular deformation in torsion

Coefficients

C_s	coefficient of elastic stability
E	modulus of elasticity of material
G	modulus of elasticity in shear
K	coefficient of specifics

Part II. Linear Theory of Infinitesimal Deformations

1. Principles of LTE

The traditional theory of elasticity is a linear theory. Within the limit of elasticity of the material this theory operates with linear equations. The domain of the elastic relation in the linear theory is considered outside of the elastic relations as the limit of elasticity of a material. In fact the elastic relation between elastic force and geometry of a body has its own limit that needs to be determined for each particular structure. Although the linear theory of elasticity has flaws, it has also very important concepts, definitions, hypotheses, and data governed by hundreds of years of intensive engineering practice and researches, which can still be used efficiently when employing the new point of view on elastic relations. Let us consider the main principles, concepts and methods of the linear theory from the angle of non-linear theory of elasticity.

The physical essence of the theory of elasticity can be described as following. External forces cause deformations and the corresponding stresses in the material of a structure. The linear theory of elasticity views a structure as built of infinitesimal elements. The whole structure in this approach is the sum of its infinitesimal parts. The theory considers stresses and strains in a small elemental volume of a structure. The total deformation, elastic force and potential elastic energy are the sums correspondingly of strains, stresses, and elastic energy of the infinitesimal parts. The linear theory of elasticity first of all considers the reversible elastic changes in a structure. After removing the forces applied to a structure the changes disappear. This theory also considers elastic deformations that are small in comparison with the overall size of a structure. Linear differential equations of stresses and strains are the basic mathematical descriptions of the theory. These linear equations do not reflect the effect of the geometry of a structure on the stress and elastic behavior of that structure. In small volumes stress is proportional to strain. This proportional relation is known as Hooke's law. This law allows one to calculate the stress in any point of a structure if the unit deformation or, in another word, strain is determined at this point analytically or experimentally. Linear theory has also a method for determining stresses that are based on knowledge of the external forces. The internal elastic forces are assumed to be in equilibrium with the external forces and distributed in accordance with the distribution of external forces. The elastic force in any section can be found as a substitute of the external forces acting on the imaginary removed part of a body.

The linear theory of elasticity has no mathematically proven descriptions of elastic relations. It is so-called empirical science. The linear theory has built its mathematical apparatus based on many hypotheses. Some of these are the following:

(1) The hypothesis of plane sections states that a section that is plane and perpendicular to the axis of a bar before deformation remains plane and perpendicular to the axis after deformation. This hypothesis is merely a convenient approximate assumption that is not always applicable to the real structure.

(2) The linear theory of elasticity considers deformations that are small in comparison with the dimensions of a structure. Hence the equations of static equilibrium of external forces are made for a structure of unchanged initial dimensions.

(3) The principle of independence of the actions of forces states:
If a structure is subjected to the action of several external forces then the total stress and deformation in a structure can be found as the sum of the deformations and stresses from the individual forces.

(4) The principle of equilibrium of internal and external forces is applied to every part of a structure. For instance, in an imaginary vertical section cutting the beam into two parts the external forces acting on either section are in equilibrium with the distributed internal forces in that section that substitute the action of the removed part. The distribution of internal forces is conveniently presented

with shear and moment diagrams in a case of bending. Thus the section with the maximum bending moment is determined. If dimensions of a beam and particularly the characteristic of a cross-section are known then corresponding maximum stress can be calculated.

This method raises doubt from the point of view of the non-linear theory of elasticity. The observation has been made that in a system in static equilibrium the forces at some small distance from the point of application cause a practically even stress distribution. The principle based on this observation is known as Saint-Venant's principle. In the literature we can find: "The state of stress in a long bar bent by couples applied to its ends is practically independent of the distribution of the forces of which the couple is the resultant." (J. Thewlis, Encyclopedic Dictionary of Physics, 1962). The point of view here is that in a system in static equilibrium the external forces acting on the body are in equilibrium with the internal forces for the system as a whole rather than for the artificially dissected parts. This hypothesis implies that the distribution of elastic forces in a body follows its own rules depending on the geometry of its structure rather than on the distribution of the external forces. The non-linear and, by the way, the linear theories of elasticity use the mean value of the external forces and bending moments in the formulas for calculating the total deformation of a structure. This point of view on the distribution of stresses is also consistent with Hooke's law for the elastic potential of a body. In addition, it is consistent with Clapeyron's formula for the potential elastic energy of a body,

$$W = \tfrac{1}{2}(\sigma_x \varepsilon_x + \sigma_y \varepsilon_y + \sigma_z \varepsilon_z + \tau_{xy}\gamma_{xy} + \tau_{yz}\gamma_{yz} + \tau_{xz}\gamma_{xz}).$$

Even though the internal forces are distributed differently from what is shown in the diagrams of distribution of external forces and moments such diagrams can be useful tools for determining the mean value of the distributed elastic forces.

(5) The hypothesis of a solid body approaches a solid body as elastic, continuous, homogeneous, and isotropic.
(6) Solid bodies are classified in the theory as bars and shells.

A bar is a body of which the length is much greater than two other dimensions. The cross-section is a section of the bar perpendicular to the axis of the bar.

A shell is a body that has one dimension much smaller than two others. This dimension is the thickness of the shell. The imaginary surface at equal distances from the outer surfaces of the shell is called middle surface. A plate has a plane middle surface.

In the following we will consider some other hypotheses.

2. Stress

External forces applied to a structure cause deformations and stresses in that structure. Stress is an internal force per unit of area. Stress at any point of a body characterizes the intensity of elastic force in the body. The vector of stress s_n in a randomly selected plane with a normal n to the plane can be resolved into components (Illustration 5). The stress perpendicular to the plane is called normal stress, σ_n, the stress parallel to the plane is called shear stress, τ_n. In the general case s_n forms some angle with the section in which it is acting. The relation between stresses is then $s_n = \sqrt{\sigma_n^2 + \tau_n^2}$. Shear stress can also be resolved in components in two perpendicular directions, $\tau_n = \sqrt{\tau_{nv}^2 + \tau_{nu}^2}$, where v and u are orthogonal axes in the n-plane.

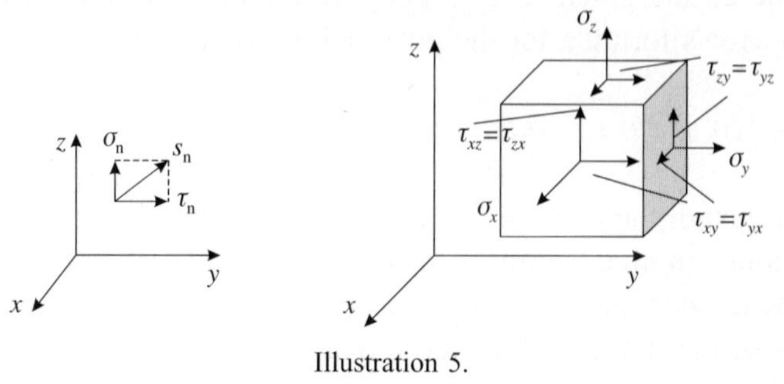

Illustration 5.

The first index for stress and strain indicates the direction of the plane, and the second index indicates the axis of projection of stress.

On the planes perpendicular to the coordinate axes x, y, z, stresses have components forming a tensor of stresses,

$$T_\sigma = \begin{Vmatrix} \sigma_x & \tau_{xy} & \tau_{xz} \\ \tau_{yx} & \sigma_y & \tau_{yz} \\ \tau_{zx} & \tau_{zy} & \sigma_z \end{Vmatrix}.$$

The stress tensor allows one to determine stress in any plane section drawn through this point, if we know components in three orthogonal planes. Only six of the components of the tensor are independent. From the condition of equilibrium of the element it follows that in orthogonal planes the shear stresses normal to the intersection of planes are equal in magnitude, $\tau_{xy} = \tau_{yx}$; $\tau_{yz} = \tau_{zy}$; $\tau_{zx} = \tau_{xz}$.

For every point of a deformed body it is always possible to find such three orthogonal planes in which the only normal stresses and shear stresses acting are equal to zero. These normal stresses are called principal stresses. The principal stresses have indexes such that $\sigma_1 \geqslant \sigma_2 \geqslant \sigma_3$. The principal stresses completely determine the stress condition at any point in a structure. If the initial planes are not the principal planes, then the principal stresses can be found as the roots of the following equation:

$$\sigma^3 - I_1\sigma^2 + I_2\sigma - I_3 = 0,$$

where

$$I_1 = \sigma_x + \sigma_y + \sigma_z;$$
$$I_2 = \sigma_x\sigma_y + \sigma_y\sigma_z + \sigma_z\sigma_x - \tau_{xy}^2 - \tau_{yz}^2 - \tau_{zx}^2;$$
$$I_3 = \sigma_x\sigma_y\sigma_z - \sigma_x\tau_{yz}^2 - \sigma_y\tau_{zx}^2 - \sigma_z\tau_{xy}^2 + 2\tau_{xy}\tau_{yz}\tau_{zx}.$$

If one of the planes is a principal plane, for example a plane perpendicular to the x-axis, and thus one of the principal stresses is known, then the other two principal stresses can be found with the formula

$$\sigma_x = \tfrac{1}{2}(\sigma_z + \sigma_y) \pm \tfrac{1}{2}\sqrt{(\sigma_z - \sigma_y)^2 + 4\tau_{yz}^2}.$$

The principal stresses determined with this formula do not depend on the known principal stress. These stresses are in the planes parallel to the x-axis. The position of those principal planes can be found with the formula

$$\tan 2\alpha_{1,2} = 2\frac{\tau_{yz}}{\sigma_z - \sigma_y}.$$

The angle α_1 is the angle between the z-axis and the normal to the principal plane with the larger of these two principal stresses; the angle α_2 is the angle between the z-axis and the normal to the principal plane with the lesser principal stress. The relation between these angles is $|\alpha_1 - \alpha_2| = 90^\circ$. The angle α should also satisfy the equation

$$\cos 2\alpha_1 = \frac{\sigma_z - \sigma_y}{\sqrt{(\sigma_z - \sigma_y)^2 + 4\tau_{yz}^2}}.$$

The angle counts from the z-axis in counterclockwise direction.

The maximum shear stress can be found as $\tau_{max} = \frac{1}{2}(\sigma_1 - \sigma_3)$, where σ_1 is the maximum principal stress, and σ_3 is the minimum principal stress in the given point.

We should note that "stress" is an unobservable theoretical concept. The magnitude of stress can be calculated but not measured. The magnitude of stress depends on our hypothesis regarding the distribution of internal forces in the material of a structure. Here we review two main hypotheses, i.e., a distribution according to Hooke's law and a distribution consistent with the hypothesis of static equilibrium of the systems of external and internal forces. Currently these two hypotheses do not appear to be in conflict. But they are. Thus, Hooke's law, $\sigma = E\varepsilon$, asserts that in an infinitesimally small volume of material, stress is proportional to strain or deformation of the infinitesimal part in consideration. The deformation of a body forms a continuous elastic line. This suggests a smooth distribution of strains in the whole body and correspondingly a smooth distribution of stresses. There might be stress-raising places in a structure, but they attribute the rise of stress to the geometry of a structure rather than to the distribution of external forces.

On the other hand, the theory of distribution of elastic forces according to the distribution of external forces projects a physical picture that obviously differs from Hooke's law. It implies sometimes-sudden changes of stresses in a structure. The distribution diagrams of forces and moments further conditioned that physical point of view. The known observations and logic of cause–effect relations corroborate with the point of view of smooth distribution of stresses in a structure, independent of the distribution of external forces applied to the structure. The exception is stress in a small part of a surface surrounding a concentrated external force.

Different hypotheses in general give different results. Thus, the stress criterion in the current approach leads in case of bending to the formula $\sigma_{max} = M_{max}/S_x \leqslant \sigma_y$, where M_{max} is the maximal external force applied to the structure. The stress criterion in case of bending with the approach of non-linear theory leads to the formula $\sigma_{max} = \sqrt{M_m E/C_s}/\mathrm{L} \leqslant \sigma_y$.

The concept of stress is important for the theory of elasticity but it is a theoretical concept. Elastic force or stress cannot be measured, therefore their objective values depend on a logical inductive connection to the observable concepts of the theory of elasticity. The magnitude and distribution of stresses can be referred to Hooke's law that connects deformation and stress, or the magnitude of elastic force can be referred to the algebraic mean value of external forces acting on a body. The external forces are the cause of deformation. They are not present in the equation of deformation that is the result of their action. In the equation of deformation we deal with theoretical elastic forces. Their presence can be explained by transforming the work of external forces into elastic energy of a body. Although the mean value of the elastic forces is equal to the mean value of the external forces, they have different origins and different rules of distribution.

3. Deformation

Deformation is a change of the dimensions and in some case shape of a body due to an applied external force. The deformation at any point is completely determined by six components of the linear deformation, ε_x, ε_y, ε_z, and the angular deformation, γ_{xy}, γ_{yz}, γ_{zx}. These components characterize linear and angular changes of the elemental parallelepiped at this point.

Linear deformation is a ratio: $\varepsilon_x = \Delta dx/dx$; $\varepsilon_y = \Delta dy/dy$; $\varepsilon_z = \Delta dz/dz$; it is considered positive for tension and negative for compression. Angular deformation is considered positive (negative) if a right angle becomes acute (obtuse). The relative volume change is $\varepsilon = \varepsilon_x + \varepsilon_y + \varepsilon_z$.

The components of deformation should also satisfy the six continuity conditions of Saint-Venant:

$$\frac{\partial^2 \varepsilon_x}{\partial y^2} + \frac{\partial^2 \varepsilon_y}{\partial x^2} = \frac{\partial^2 \gamma_{xy}}{\partial x \partial y}, \quad \frac{\partial^2 \varepsilon_y}{\partial z^2} + \frac{\partial^2 \varepsilon_z}{\partial y^2} = \frac{\partial^2 \gamma_{yz}}{\partial y \partial z}, \quad \frac{\partial^2 \varepsilon_z}{\partial x^2} + \frac{\partial^2 \varepsilon_x}{\partial z^2} = \frac{\partial^2 \gamma_{xz}}{\partial x \partial z},$$

$$\frac{\partial^2 \varepsilon_x}{\partial y \partial z} = \frac{1}{2} \frac{\partial}{\partial x} \left(\frac{\partial \gamma_{xy}}{\partial z} + \frac{\partial \gamma_{xz}}{\partial y} - \frac{\partial \gamma_{yz}}{\partial x} \right),$$

$$\frac{\partial^2 \varepsilon_y}{\partial x \partial z} = \frac{1}{2} \frac{\partial}{\partial y} \left(\frac{\partial \gamma_{yz}}{\partial x} + \frac{\partial \gamma_{xy}}{\partial z} - \frac{\partial \gamma_{xz}}{\partial y} \right),$$

$$\frac{\partial^2 \varepsilon_z}{\partial x \partial y} = \frac{1}{2} \frac{\partial}{\partial z} \left(\frac{\partial \gamma_{xz}}{\partial y} + \frac{\partial \gamma_{yz}}{\partial x} - \frac{\partial \gamma_{xy}}{\partial z} \right).$$

This mathematical construct is based on the assumed knowledge of deformation in infinitesimal volume of the material of a structure. If successful, then after all these calculations for determining the deformation, the linear theory has no criteria for evaluating it. If we are to evaluate the deformation we need to compare rates of change of deformation, which linear theory cannot provide.

Deformation of a body causes linear and angular displacements relative to the initial position of the body. The complete linear displacement, δ, can be expressed in its projections on the coordinate axes, $\delta = \sqrt{u^2 + \upsilon^2 + w^2}$, where u is the displacement in the direction of the x-axis, υ is the displacement in the direction of the y-axis, and w is the displacement in the direction of the z-axis.

In most cases the deformations and displacements are small in comparison with the dimensions of a body. However, some structures such as springs and membranes have large displacements corresponding to small deformations.

The non-deformed part of a body can be displaced as well. The displacement depends on the design of a structure. For stress analysis however only the stresses in the deformed part of a structure are important. The criterion for stress analysis selected in the linear theory of elasticity is the limiting characteristic of the material. In engineering practice it is recognized that in most cases the real limit for a structure is different from the limiting stress of the material. In an attempt to adjust the limit the effect of each design factor is studied and the limit is manipulated with those factors. This is the method empirical sciences usually use in order to compensate for theoretical inadequacies.

4. Hooke's Law

The linear theory of elasticity considers unit deformations as the basis for describing elastic changes in a body. The unit deformation or strain is proportional to the unit stress in a body. This relation is known as Hooke's law. The ratio of stress and strain is a constant for a given material. The coefficient E is called the modulus of elasticity and G is called the shear modulus. The modulus of elasticity is the ratio of normal stress to the unit of linear deformation, $\sigma/\varepsilon = E$. The shear modulus is the ratio of shear stress to the angular deformation, $\tau/\gamma = G$. The relation between modulus of elasticity and shear modulus is $E = 2G(1+\mu)$, where μ is Poisson's ratio, which characterizes the contraction of cross-sectional dimensions with elongation of the longitudinal dimension. For example, for uniaxial tension if $\varepsilon_z = \sigma_z/E$ then $\varepsilon_x = \varepsilon_y = \mu\varepsilon_z = \mu\sigma_z/E$. For an isotropic body Hooke's law provides the following equations:

$$\varepsilon_x = \frac{1}{E}[\sigma_x - \mu(\sigma_y + \sigma_z)]; \qquad \gamma_{xy} = \frac{1}{G}\tau_{xy};$$

$$\varepsilon_y = \frac{1}{E}[\sigma_y - \mu(\sigma_x + \sigma_z)]; \qquad \gamma_{yz} = \frac{1}{G}\tau_{yz};$$

$$\varepsilon_z = \frac{1}{E}[\sigma_z - \mu(\sigma_x + \sigma_y)]; \qquad \gamma_{xz} = \frac{1}{G}\tau_{xz}.$$

The external forces acting on a body perform the work of deformation. This work transforms into potential energy of the deformed body. The unit of elastic potential energy can be determined with the equation of Clapeyron,

$$w = \tfrac{1}{2}(\sigma_x\varepsilon_x + \sigma_y\varepsilon_y + \sigma_z\varepsilon_z + \tau_{xy}\gamma_{xy} + \tau_{yz}\gamma_{yz} + \tau_{zx}\gamma_{zx}).$$

Stress condition	Relations
Uniaxial stress 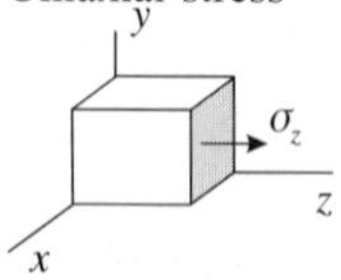 	Components of deformation: $\varepsilon_z = \sigma_z/E,\ \varepsilon_y = \varepsilon_x = -\mu\sigma_z/E;\ \gamma_{xy} = \gamma_{yz} = \gamma_{zx} = 0.$ Components of stress: $\sigma_z = E\varepsilon_z;\ \sigma_y = \sigma_x = 0;\ \tau_{xy} = \tau_{yz} = \tau_{zx} = 0.$ Limits for stress and strain: $\varepsilon_z \leqslant AC_s/L^2 \leqslant \sigma_{cr}/e;\ \sigma_z \leqslant AC_sE/L^2 \leqslant \sigma_y.$
Shear stress 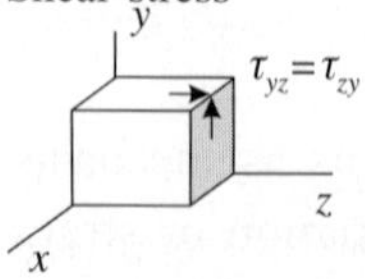 	Components of deformation: $\varepsilon_x = \varepsilon_y = \varepsilon_z = 0;\ \gamma_{xy} = \gamma_{zx} = 0;\ \gamma_{yz} = \tau_{yz}/G.$ Components of stress: $\sigma_x = \sigma_y = \sigma_z = 0;\ \tau_{xy} = \tau_{xz} = 0;$ $\tau_{zy} = G\gamma_{yz} = E\gamma_{yz}/[2/(1+\mu)].$ Limit for stress: $\tau_{zy} \leqslant AGC_s/L^2 \leqslant \tau_y.$
Biaxial stress 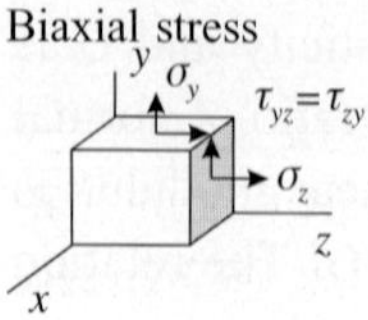 	Components of deformation: $\varepsilon_x = -\mu(\sigma_y + \sigma_z)/E;\ \varepsilon_y = (\sigma_y - \mu\sigma_z)/E;\ \varepsilon_z = (\sigma_z - \mu\sigma_y)/E;$ $\gamma_{xy} = \gamma_{zx} = 0;\ \gamma_{yz} = \tau_{yz}/G.$ Components of stress: $\sigma_x = 0;\ \sigma_y = E(\varepsilon_y + \mu\varepsilon_z)/(1-\mu^2);$ $\sigma_z = E(\varepsilon_z + \mu\varepsilon_y)/(1-\mu^2);$ $\tau_{yz} = G\gamma_{yz} = E\gamma_{yz}/2(1+\mu).$ Limit for stress and strain: $\sigma_1 = AC_sE/L^2 < \sigma_{cr};\ \varepsilon_1 = AC_s/L^2 < \sigma_{cr}/E.$
Triaxial stress 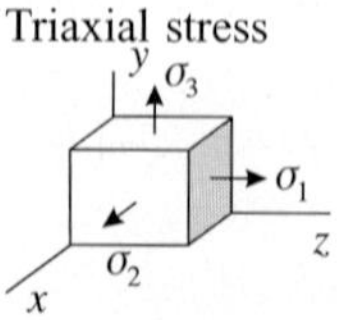 	Components of deformation: $\varepsilon_1 = [\sigma_1 - \mu(\sigma_2 + \sigma_3)]/E;\ \varepsilon_2 = [\sigma_2 - \mu(\sigma_3 + \sigma_1)]/E;$ $\varepsilon_3 = [\sigma_3 - \mu(\sigma_1 + \sigma_2)]/E.$ Limit of stress and strain: $\sigma_1 = AC_sE/L^2 < \sigma_{cr},\ \varepsilon_1 = AC_s/L^2 < \sigma_{cr}/E.$
Triaxial stress 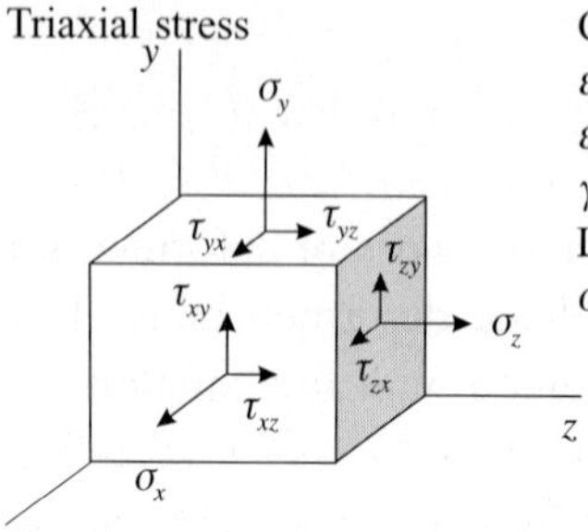 	Components of deformation: $\varepsilon_x = [\sigma_x - \mu(\sigma_y + \sigma_z)]/E;\ \varepsilon_y = [\sigma_y - \mu(\sigma_z + \sigma_x)]/E;$ $\varepsilon_z = [\sigma_z - \mu(\sigma_x + \sigma_y)]/E;$ $\gamma_{xy} = \tau_{xy}/G;\ \gamma_{yz} = \tau_{yz}/G;\ \gamma_{zx} = \tau_{zx}/G.$ Limit for stress and strain; $\sigma_1 = AC_sE/L^2 < \sigma_{cr},\ \varepsilon_1 = AC_s/L^2 < \sigma_{cr}/E.$

Potential elastic energy can be expressed with the stress components,

$$w = \frac{1}{2E}\Big[\sigma_x^2 + \sigma_y^2 + \sigma_z^2 - 2\mu(\sigma_x\sigma_y + \sigma_y\sigma_z + \sigma_z\sigma_x) + 2(1+\mu)(\tau_{xy}^2 + \tau_{yz}^2 + \tau_{zx}^2)\Big].$$

Potential elastic energy can also be expressed with the strain components,

$$w = G\left[\varepsilon_x^2 + \varepsilon_y^2 + \varepsilon_z^2 + \frac{\mu}{1-2\mu}(\varepsilon_x + \varepsilon_y + \varepsilon_z)^2 + \tfrac{1}{2}(\gamma_{xy}^2 + \gamma_{yz}^2 + \gamma_{zx}^2)\right].$$

The total elastic energy of a body is found by integrating over the volume V of the body: $W = \int w\,\mathrm{d}V$. The volume change can be calculated with the equation $\varepsilon = 3k\sigma$, where $\sigma = \frac{1}{3}(\sigma_x + \sigma_y + \sigma_z)$, with k the coefficient of volume contraction, $k = (1-2\mu)/E$. Then, $w = \frac{3}{2}k\sigma^2 + \tau_i/2G$, where τ_i is the intensity of shear stress, $\tau_i \approx 1.08\tau_{max}$.

The important theorems referring to the elastic energy are:

Clapeyron Theorem: The potential energy of a body is equal to one half of the product of the external forces and the displacements cause by these forces, $\int w\,\mathrm{d}V = \frac{1}{2}[\int(Xu + Yv + Zw)\,\mathrm{d}V + \int(X_n u + Y_n v + Z_n w)\,\mathrm{d}S$, where S is the surface of the body, and X_n, Y_n and Z_n are the forces at its surface.

Principle of minimum potential energy of system: Of all the kinetically possible systems of displacements at the surface of a body, the actual displacements correspond to the minimal potential energy of that system.

5. Geometric characteristics of plane areas

For the calculation of stresses other than those for familiar geometrical characteristic such as area of cross-section, additional cross-sectional characteristics were introduced in the mechanics of rigid bodies and used in the mechanics of elastic bodies. These characteristics allow one to calculate the stresses corresponding to bending or torsion moments. These geometrical characteristics are also used for calculating the geometrical stiffness of a structure for different stress–deformation conditions.

The sectional area can be found with the integral $A=\int \mathrm{d}A$. The static moments of this area with respect to the axes are correspondingly $S_x=\int y\,\mathrm{d}A$, $S_y=\int x\,\mathrm{d}A$. The moments of inertia of the cross-section with respect to the axes are correspondingly $I_x=\int y^2\,\mathrm{d}A$, $I_y=\int x^2\,\mathrm{d}A$; the section modulus and moment of inertia can be calculated with the formula $S_x=I_x/y_{\max}$; the product of inertia can be found with the integral $I_{xy}=\int xy\,\mathrm{d}A$. The polar moment of inertia, $I_\mathrm{p}=\int \rho^2\,\mathrm{d}A$ or $I_\mathrm{p}=I_x+I_y$, is used for determining stress in torsion; the polar section modulus is $S_\mathrm{p}=I_\mathrm{p}/c$. Tables of moments of inertia and section modulus for common cross-sections are found in all handbooks for mechanical and civil engineers.

If a body has several parts then its moment of inertia is equal to the sum of the moments of inertia of its parts with respect to the same axis. The moment of inertia can also be presented as $I_x=\int y^2\,\mathrm{d}A=k^2A$, where k is defined as the radius of gyration.

The relation between moments of inertia with respect to the axes x' and y' at an angle α to the axes x and y are:

$$I_{x'} = I_x \cos^2\alpha + I_y \sin^2\alpha - I_{xy}\sin 2\alpha;$$
$$I_{y'} = I_x \sin^2\alpha + I_y \cos^2\alpha + I_{xy}\sin 2\alpha.$$

The principal moments of inertia are the moments of inertia with respect to the principal axes; these are orthogonal axes that have product of inertia zero. The principal moment of inertia with respect to one axis has maximum magnitude and the principal moment of inertia with respect to the other axis has minimum value.

The magnitudes of the principal moments of inertia can be found with the formulas

$$I_{\max} = \tfrac{1}{2}(I_x + I_y) + \sqrt{\tfrac{1}{4}(I_x - I_y)^2 + I_{xy}^2},$$
$$I_{\min} = \tfrac{1}{2}(I_x + I_y) - \sqrt{\tfrac{1}{4}(I_x - I_y)^2 + I_{xy}^2},$$

where I_x, I_y and I_{xy} are the moments of inertia about the initial axes.

The principal axes that are drawn through the center of gravity of a section are called principal central axes. For a symmetrically shaped section one of the principal central axes is the axis of symmetry and the other is the central axis perpendicular to it. For an asymmetrically shaped section the position of the principal axes can be found with the formula $\tan 2\alpha = 2I_{xy}/(I_y - I_x)$.

The geometrical characteristics of the cross-section of a bar are used for the determining the stress in the cross-section. In case of tension–compression, the principal stress normal to the cross-section is $\sigma = N/A$ lb/in^2 (or kg/cm^2), where N (lb or kg) is the mean value of the elastic normal to the cross-section force and A (in^2 or cm^2) is the cross-sectional area. In case of bending the mean value of principal normal stress is $\sigma = M/S_x$ lb/in^2 (or kg/cm^2) where M (in-lb or cm-kg) is the mean value of the elastic moment in the cross-section and S (in^3 or cm^3) is the section modulus. Thus these geometrical characteristics are designed to serve our purpose of determining the stress per unit of an area.

According to the current theory of strength, the strength of a structure depends on the maximum stress in the structure that should be less than the limiting mechanical properties of the material.

6. Combination of stresses

In case of combination of stresses the stress condition is reduced to the equivalent principal stresses. There are strength hypotheses for such cases.

(1) In the energy theory the stress systems are considered to be equivalent if they have equal specific potential energy of deformation (elastic energy attained by the unit volume of a body). The equivalent stress is determined with the equations

$$\sigma_{eq} = \sqrt{\sigma_x^2 + \sigma_y^2 + \sigma_z^2 - (\sigma_x\sigma_y + \sigma_y\sigma_z + \sigma_z\sigma_x) + 3(\tau_{xy}^2 + \tau_{yz}^2 + \tau_{zx}^2)} \quad \text{or}$$

$$\sigma_{eq} = \sqrt{\sigma_1^2 + \sigma_2^2 + \sigma_3^2 - (\sigma_1\sigma_2 + \sigma_2\sigma_3 + \sigma_3\sigma_1)}.$$

(2) The theory of maximum shear stresses regards two stress conditions as equivalent if they have equal maximum shear stresses. The formula for equivalent stress in this theory is $\sigma_{eq} = \sigma_1 - \sigma_3$.

(3) Mohr's strength theory is used for calculating equivalent stress when the material has different strength characteristics in tension and in compression. Then the formula of equivalent stress is $\sigma_{eq} = \sigma_1 - \mu\sigma_3$.

In the linear theory of elasticity we have equations which operate with stresses and strains related to each other according to Hooke's law. It is understood that stress can be determined because we know the magnitudes and distribution of the internal forces, which are in equilibrium with the external forces. From the condition of static equilibrium the generalized form of Hooke's law provides equations for the different cases that are considered in courses on the strength of materials. Unknown components are determined from such equations. For example, according to Hooke's law $\varepsilon = \sigma/E$.

The total normal stress in a cross-section of a bar is $\sigma = N/A - yM_x/I_x + xM_y/I_y$. Mathematics inconspicuously has led us to different equations, which include additional hypotheses. The important hypothesis that appears here is that the elastic relation for an infinitesimal unit can be extended to the whole structure.

The main purpose of the linear theory of elasticity is stress analysis. It is assumed that determining maximum principal stress allows one to evaluate the strength of a structure. There are different theories of strength, which will be discussed later. However, from our discussion in Part I the analysis of stresses and the limiting stress criterion are insufficient for evaluating the strength and elastic stability of a structure. The limiting strength characteristic for a structure is not some fixed value of stress or strain. The structure has an individual limiting stress and strain at failure; its elastic behavior and failure are characterized by the rate of change of deformation rather than with some fixed value of stress. The stress analysis can be conducted only after desirable optimal dimensions of a structure are established. If the stress in such a structure exceeds the limit of the selected material then the cross-sectional dimensions can be increased or the engineer may select a stronger material. If, on the other hand, the limit of the structure generated by its geometry is much less than the limit of the material and we want to make better use of the elastic properties of the material we also have several options. We may select a material with a lower limit, or we can change the geometry of the structure in the direction of reducing its excessive strength characteristic. However, we should keep in mind that stress analysis by itself is insufficient as the only tool for structural design. A quantitative law describing the relative change of elastic behavior of the structure is needed for the proper evaluation of both the limiting stress and the limiting rate of change of deformation that characterizes stable elastic behavior of the structure. The current engineering stress analysis practically ignores the effect of geometry on deformation and stress. For example, consider similar beams made of the same material and subjected to the action of the same bending moment. If the beams have the same cross-sections but different lengths, according to the formula for stress these beams have the same stress and the same allowable stress, $\sigma = M/S_x \leqslant \sigma_a$. The effect of the length of the beam and subsequently the effect of deformation on the strength and stability of the beam is neglected against common observations and reason. The scientific reason for taking into account deformation for establishing the individual limiting stress was discussed in Part I. The individual limit of elasticity of a structure can be calculated with the differential equation of deformation in the interval of rapid increase of deformation and failure of elastic behavior:

$$D = \frac{M}{ER}; \qquad \frac{\mathrm{d}D}{\mathrm{d}R} = -\frac{M}{ER^2}; \qquad \frac{M}{ER^2} = C_s, \quad C_s = [\tan \alpha].$$

This is a comprehensive mathematical and empirical law.

6.1 Load and Resistance Factor Design (LRFD)

Since 1994 a standard has been created based on a Two-Limit-States design procedure. First, the strength criterion is compared with the resistance criterion. The method is said to be based on probability approach. The load effect on strength and the resistance effect are compared. "Theoretically, the structure will not fail unless *R* is less than *Q*." (The Civil Engineering Handbook, Editor-in-Chief W.F. Chen, 1995).

There is no theoretical foundation for this method, though the Civil Engineering Handbook predicts "Allowable stress design has been the norm of the design profession for decades. Nevertheless, the worldwide trend is toward the limit state approach to design. This change in design philosophy is evidenced by the adaption and general usage of various limit state specifications in Canada, Europe and Japan. In view of this trend and cognizance of the likelihood that LRFD will be the mainstream design method in the 21st century, only LRFD provisions will be discussed in this chapter." (The Civil Engineering Handbook, Editor-in-Chief W.F. Chen, 1995). In the Manual of Steel Construction, vol. 1, 1994, we read the following remark concerning the theory of LRFD: "The design philosophy for tension members is the same in the LRFD and ASD [allowable stress design] specifications. Although the column strength equations have been revised for compatibility with LRFD ... the philosophy and procedures of column design in LRFD are similar with those in ASD. ... [LRFD] is a method of proportioning structural members using load factors and resistance factors such that no limit state is exceeded." (Manual of Steel Construction, AISC, 1994). As a matter of fact, no method existed in the engineering literature that would compare structures in order to select the structure with the optimal correlation of longitudinal and cross-sectional dimensions. For comparing structures one should have a theoretical criterion of comparison. There is no such criterion and no explanation is given in LRFD how to reach similarity of structures. The author's method of optimal design does have a theoretical criterion for comparing structures and for establishing the optimal correlation between cross-sectional and longitudinal dimension. I proposed my method to AISC in 1981. The concept of geometrical stiffness first appeared in my work in 1986.

I have proposed theoretically comprehensive two-limit design procedures, known to some engineering institutions such as AISC, NSF, ASI, ASC&ME since 1981, as well copyrighted the articles many times following the development of my method. I published an article "Deflections indicate the Design Similarity" in "Machine Design" in 1986, patented "Method of and Apparatus for Optimization of Structures" in 1997, and published an article "Method of Optimization the Structures" in 1999. The method of optimization and NLTE are a new comprehensive physical theory and a new method based on this theory. The LRFD method, on the other hand, refers to some statistical data that is not in the literature and has no scientific explanation.

Part III. Optimization of typical structures

1. Introduction

The problems concerning structure design and the method of optimization of a structure have been described in general terms in Part I. In Part III we will be dealing with the calculation of the optimal dimensions for typical structures. The purpose, however, is the same as in the Part I, that is, to provide a reliable approach to the design of structures. The examples should be understood as exemplary teachings only.

Optimal design requires a reliable description of elastic relations, a reliable criterion for optimizing a structure, knowledge of the stress distribution, knowledge of the individual limiting stress, and knowledge of the limiting deformation. Here we will consider optimization of typical structures subjected to tension, compression, shear, torsion, bending, and their combinations.

In order to find a structure-specific limit and to establish the optimal dimensions within this limit we use our new equation for elastic deformation. This equation describes the behavior of a family of similar structures. The members of the family are distinguished from each other by their resistance to deformation, *ER*. The important variable part of the resistance is a characteristic defined as geometrical stiffness, *R*. Graphical presentation of the elastic relation, such as deformation versus geometrical stiffness, gives an accurate comprehensive picture for the selection of a structure of optimal geometrical stiffness. Geometrical stiffness is a geometrical characteristic of a structure as it relates to the specific deformation. For a bar in tension, geometrical stiffness is $R=KA/L$; For a bar subjected to torsion it is $R=KI_p/L$; and for a bar in bending it is $R=KI/L$. The coefficient K accounts for the effect of all known and unknown specific features of design on the geometrical stiffness that are not included otherwise in the formula. The geometrical stiffness of a real structure can be calculated if K is known or measured if K is unknown.

Our fundamental approach to the problems of engineering design is provided with a mathematical description of the elastic relations. This description consists of two interconnected equations. The basic equation establishes relations among the variables, namely deformation, elastic force, geometrical stiffness and modulus of elasticity of material. The deformation is proportional to the mean value of the distributed elastic force and inverse to the geometrical stiffness and modulus of elasticity, $D=F/ER$. For selecting an optimal geometrical stiffness we have to evaluate the effect of the geometrical stiffness on deformation. This evaluation can be made with the derivative equation. The derivative equation describes the rate of change of deformation due to change of geometrical stiffness, $\mathrm{d}D/\mathrm{d}R=-F/ER^2$. This equation is called the equation of elastic stability. The equation of elastic stability allows us to select a structure with mathematically certain elastic behavior. It also allows us to calculate the structure-specific elastic limit.

The limit of elasticity is characterized by a rapid increase of deformation. For a given force, the limit can be calculated corresponding to the critical value of geometrical stiffness. On a plot of deformation versus geometrical stiffness the rate of deformation at any point corresponds to the tangent at this point. For steel structures the rate corresponding to elastic failure has the approximate value $\tan 76^\circ=4$. In the equation of elastic stability the particular rate is represented with the coefficient of elastic stability, $C_s=|\tan\alpha|$. Note that in no case may the rate exceed the approximate value $\tan 76^\circ=4$.

The behavior of a structure depends on the position of that structure in a set of similar structures distinguished by their geometrical stiffness. The equation that describes the behavior is the differential equation $\mathrm{d}D/\mathrm{d}R=-F/ER^2$. For any concrete structure the equation takes the form $F/ER^2=C_s$, where C_s is the selected desirable value of the rate.

The system of equations $D=F/ER$ and $F/ER^2=C_s$ allows us to solve the basic engineering problems. The selection of the coefficient of elastic stability C_s predicts the behavior of a structure, i.e. how stable a particular structure will be, and how rapid deformation would increase/decrease with fluctuations in the dimensions and fluctuations in the force acting on the structure. By selecting the critical value $C_s = 4$ we may calculate the critical cross-sectional characteristics A_{cr} or S_{cr} corresponding to the critical geometrical stiffness R_{cr}, from $R_{cr} = KA_{cr}/L$ or $R_{cr} = KS_{cr}/L$. And thus we can calculate the limiting stress for a given internal force F, $\sigma_{cr} = N/A_{cr}$ or $\sigma_{cr} = M/S_{cr}$.

Geometrical stiffness can be expressed mathematically as the ratio $R = KA/L$ in the case of tension, or $R = KI/L$ in the case of bending, where I is the moment of inertia of the cross-section. Geometrical stiffness can be calculated with the equation of elastic stability, $R = \sqrt{F/EC_s}$. The coefficient of elastic stability is selected with the knowledge of the character of the function of elastic deformation and the specific working conditions for a structure, in particular, the knowledge of the forces applied to the structure. For most practical cases the value $C_s = 1$ ($\tan 45^o$) gives a sufficient factor of safety, $n = \sigma_{cr}/\sigma_{max} = 2$. The knowledge of the actual limit of a structure and the selection of a desirable rate of change of deformation give us the opportunity to select a factor of safety accordingly.

The variables in the system of equations were selected in such a way that the function of elastic stability has the character of a tangent function. Like a tangent function it has two intervals: the interval of rapid changes, $1 < C_s < 4$, and the interval of slow changes, $0 < C_s < 1$. Any point of the curve D vs. R has a coefficient of elastic stability corresponding to the tangent at this point. The coefficient of elastic stability C_s can be selected in the interval of the rapid changes or in the interval of slow changes depending on our purposes.

The equations $R = KA/L$ and $R = KI/L$ both contain K, the coefficient of design specifics. This coefficient accounts for the effect of irregularities of cross-section, such as holes, notches, change of cross-section and other stress raisers. It also accounts for the effect of the boundary conditions, particularly the effect of the supports on the geometrical stiffness of a structure. This coefficient of specifics can be obtained experimentally, by measuring geometrical stiffness of a structure, $R = N/Ee$, cross-sectional dimension and length, $K = RL/A$. In simple cases it can be included in the reduced characteristic of a cross-section. For a uniform cross-section, $K = 1$, and then $A = RL$.

Geometrical stiffness in the theory of elasticity is a comprehensive physical characteristic that shows the resistance of a structure to deformation due to the geometry of a structure. Geometrical stiffness has a mathematical description. Two similar structures may have different geometrical stiffness. For structures of equal length, the resistance to deformation is greater for the structure with a greater cross-sectional area. For structures of equal cross-section, the resistance is less for the longer structure. However, for more complex structures such as shells, and in the case of combined stresses, it is a more difficult task to adequately describe geometrical stiffness; it is much easier to measure it. After calculating the optimal geometrical stiffness the optimization of a particular structure can be achieved by changing its cross-section, or the longitudinal dimension, or any other geometrical parameter like its radius of curvature, if its effect on the geometrical stiffness is known.

In the following we will apply the method of optimization described in general in Part I to specific structures and particular cases. The fundamental equations for the structural design are the equation of elastic deformation, $D = F/ER$ and the equation of elastic stability, $F/ER^2 = C_s$.

Some general principles concerning stress and strain distribution and the relation between stress and strain are common to the linear and non-linear theories of elasticity. Some other principles, accounting for the effect of the geometry on the elastic behavior, are added.

(1) The linear theory deals with the description of an elastic line representing a physical body.

(2) The non-linear theory of elasticity describes the mathematical relations in a family of similar structures. The elastic behavior of an individual structure depends on its place in the family.

(3) In the linear theory of elasticity, the stress–strain relation and the limit of elasticity of the material form the basis for structural analysis and design. Thus, structures made of the same material and subjected to the same equal external forces will, if they have the same cross-sectional characteristics, have the same stresses, these being the design criterion in the linear theory. The effect of length on the resistance to deformation is thus not adequately considered as a factor of design.

(4) In the non-linear theory the description of elastic relations for the whole body and the rate of change of deformation due to

geometrical stiffness of the body form the basis for structural analysis and design. The calculations are based on the non-linear function describing deformation, elastic force, geometrical stiffness and modulus of elasticity relations, $D = F/ER$, and the derivative function describing the rate of change of deformation due to the geometrical stiffness of the structure, $F/ER^2 = C_s$.

(5) On the infinitesimal level of stress–strain relation any deformation s in a small volume can be presented as an elongation in three perpendicular main directions, $ds^2 = dx^2 + dy^2 + dz^2$.

(6) For a small volume there is a function W, called the elastic potential, which describes relations, $\sigma_x = dW/d\varepsilon_x$, $\sigma_y = dW/d\varepsilon_y$, $\sigma_z = dW/d\varepsilon_z$; $\tau_{xz} = dW/d\gamma_{xz}$, $\tau_{xy} = dW/d\gamma_{xy}$, $\tau_{yz} = dW/d\gamma_{yz}$. Alternatively, $\varepsilon_x = dW/d\sigma_x$; $\varepsilon_y = dW/d\sigma_y$; $\varepsilon_z = dW/d\sigma_z$; $\gamma_{xz} = dW/d\tau_{xz}$, $\gamma_{xy} = dW/d\tau_{xy}$, $\gamma_{yz} = dW/d\tau_{yz}$.

Hooke's law in the linear and non-linear theory assumes the linear correlation between the stress components and the strain components.

(7) The elastic potential energy for the unit volume can be obtained with the formulae

$$W = \tfrac{1}{2}E\Big[\sigma_x^2 + \sigma_y^2 + \sigma_z^2 - 2\mu(\sigma_x\sigma_y + \sigma_y\sigma_z + \sigma_x\sigma_z) + 2(1+\mu)(\tau_{xy}^2 + \tau_{yz}^2 + \tau_{xz}^2)\Big],$$

$$W = G\left[\varepsilon_x^2 + \varepsilon_y^2 + \varepsilon_z^2 + \frac{\mu\varepsilon^2}{1-2\mu} + \tfrac{1}{2}(\gamma_{xy}^2 + \gamma_{yz}^2 + \gamma_{xz}^2)\right].$$

(8) For solving an engineering problem we should know the distribution of stress in the structure. One hypothesis of stress distribution is Saint-Venant's principle. It describes the distribution of stresses on the surface of a structure. At small distances from the applied force the stresses are distributed uniformly.

(9) In the non-linear theory the elastic force F and the elastic moment M in the formulae for stress and deformation are the mean values of the forces distributed uniformly in a structure of uniform cross-section.

(10) The potential elastic energy of a body is a product of the internal force on the total deformation. For tension it is $U = Ne/2$; for torsion it is $U = M_T\theta/2$; and for bending it is $U = M\theta/2$. For a bar having parts with different geometrical stiffnesses the total potential energy of deformation is equal to the algebraic sum of the potential energy of the parts, $U = \sum U_m$, where m is the number of parts.

(11) For a bar having m parts the total elongation is equal to the algebraic sum of the deformations of the parts.

(12) The total geometrical stiffness of a bar having parts with different geometrical stiffnesses is equal to the product of the values of geometrical stiffness divided by their sum. For example, in tension the total geometrical stiffness is $R = R_1R_2/(R_1 + R_2)$. (Note the similarity and the difference between Ohm's law $I = V/R$ where $R = \rho L/A$, and the law of elasticity, $D = F/ER$ where $R = KA/L$.)

(13) The assumption in the linear theory is that deformation in tension is proportional to the normal stress and the shear deformation is proportional to the shear stress.

(14) An assumption in this work for the stress distribution, particularly in torsion, is that angular deformation is proportional to the product of shear stress and the distance of a point from the axis. Also, shear stress is distributed uniformly in a uniform cross-section.

(15) An assumption in common with the linear theory of elasticity is that a cross-section which is plane and perpendicular to the axis of a bar before deformation remains plane and perpendicular to the axis after deformation. However, this assumption is not applicable to the torsion of non-round bars.

(16) The calculation of dimensions for a structure under dynamic stresses in the non-linear theory of elasticity includes both knowledge of the limiting stress of the material and knowledge of the critical stress generated by the geometry of the structure.

(17) The elastic deformations, stresses and displacements of a structure do not depend on the order of loading but are defined by the end loading condition. This implies the principle of independent action of forces. The deformations and displacements induced by multiple forces equal the sums of the deformations or displacements from the individual forces acting independently. The principle of

independence is not applicable to the action of forces that change the character of deformation.

(18) In the general case the equivalent elastic force can be calculated as the sum of the force vectors. The shear force in the cross-section is $Q=\sqrt{Q_x^2+Q_y^2}$, and the elastic moment is $M=\sqrt{M_x^2+M_y^2}$.

2. Tension/compression

The main assumptions we have made for a bar in tension/compression are as following: the straight axis of the bar remains straight after deformation, cross-sections perpendicular to the axis before deformation remain perpendicular after deformation. For a bar of uniform cross-section the assumption is that stresses are distributed evenly along the bar under stress. For materials having the same modulus of elasticity for tension and compression the deformations are distinguished by their sign: by convention, a positive value denotes tension, a negative value compression.

The basic system of equations for a bar in tension is $e=N/ER$ and $N/ER^2=C_s$. The total deformation of the bar is e. N, the internal force normal to the cross-section, is equal to the sum of the normal stresses uniformly distributed in the cross-section. The geometrical stiffness of a bar of uniform cross-section is the ratio $R=A/L$. Optimal geometrical stiffness can be calculated with the equation for elastic stability, $R_o=\sqrt{N/EC_s}$. If the selected coefficient $C_s=1$ (tan 45°), then $R_o=\sqrt{N/E}$. If the length of a bar is L then the optimal area for the cross-section is $A_o=R_oL$. For determining the actual limit of elasticity of the bar we have to compare the individual limit obtained with the equation of elastic stability with the limit of the material obtained experimentally. The limit depending on the geometry of a structure required by the elastic force N corresponds to $R_{cr}=\sqrt{N/EC_s}$. The coefficient of elastic stability at the critical point is $C_s=4$ (tan 76°). The area of the cross-section at this point is $A_{cr}=R_{cr}L$. The individual limit of elasticity is $\sigma_{cr}=N/A_{cr}$. If $\sigma_{cr}<\sigma_y$ we include a safety factor in calculating the cross-sectional area. Thus, $C_s=1$ corresponds to a safety factor of $n=2$, $\sigma_o=\frac{1}{2}\sigma_{cr}$. For $C_s=\tan 60°=1.73$, the safety factor is $n=\sqrt{4/1.73}=1.5$.

If the yield stress of the material is less than the calculated critical stress, $\sigma_y < \sigma_{cr}$, we have to calculate the cross-sectional area based on the strength-of-material approach, $A_{cr} = N/\sigma_y$, and apply the safety factor to this value. The allowable stress in this case is $\sigma_a = \sigma_y/n \leqslant \sigma_{cr}$.

The actual elastic limit of the bar is the lesser of the two values of limiting stress, either σ_y or σ_{cr}. This knowledge of the actual limit of a structure makes the calculation of dimensions far more reliable and significantly reduces the necessary safety factor. The total deformation of a bar of optimal geometrical stiffness can be calculated, $e = N/ER_o$.

If the actual force acting in a structure is unknown, then the elastic force distributed in the structure can be calculated by measuring the deformation from the actual forces, $N = EeR_a$.

The unit deformation of cross-section ε' in tension/compression can be calculated using the known relation $\varepsilon' = -\mu\varepsilon$, where μ is Poisson's ratio and $\varepsilon = e/L$.

The geometrical stiffness of a bar of more complex design, characterized by a variable cross-section, can be measured by measuring the total elongation of the bar. Any experimental force N allows one to find the actual geometrical stiffness of the structure, $R_a = N/Ee$. Also, the total geometrical stiffness of a bar having parts of different geometrical stiffness can be calculated. The total deformation of such a bar is equal to the sum of the deformations of its parts. For example: a bar loaded with axial force N and consisting of two parts has total deformation $e = e_1 + e_2$; or $N/ER = N/ER_1 + N/ER_2$; after transformation, $1/R = 1/R_1 + 1/R_2$ or $R = R_1R_2/(R_1 + R_2)$.

The elastic forces acting in a structure can be calculated if the external forces are known. If the equations of statics, $\sum F = 0$ and $\sum M = 0$, are sufficient for determining elastic forces, then the problem is considered as statically determinate. If the number of parts (m) of a complex structure and the number of forces acting in these parts exceeds the number of equations available from statics, then the system is statically indeterminate. For calculating the internal forces for such system we construct, usually, the necessary additional equations of deformations.

Note that in all cases for a statically determinate or indeterminate system the elastic forces can be found by a non-destructive test measurement of deformations of the parts. After that the force in each part can be calculated, $N = eER_a$.

In the following we present some examples of calculating the optimal cross-sectional dimensions of the bars.

Example 1: Statically determinate structure

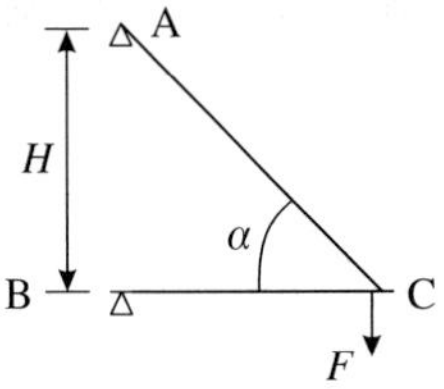

$F = 6{,}000$ lbs, H = 30 in, $L = AC = 60$ in, $E = 30\times10^6$ psi, $\sigma = 30{,}000$ psi, $C_s = 1$, $\alpha = 30°$.
$N = F/\sin\alpha = FL/H = 12{,}000$ lbs;
$R_o = \sqrt{N/EC_s} = \sqrt{12{,}000/30\times10^6} = 0.02$ in; $A_o = R_oL = 0.02*60 = 1.2$ in^2.
For $C_s = 1$, the safety factor is $n = 2$.
Stress in the structure is $\sigma_o = N/A_o = 12{,}000/1.2 = 10{,}000$ psi.
For $C_s = 4$, $R_{cr} = 0.01$ in, $A_{cr} = 0.6$ in^2, $\sigma_{cr} = 20{,}000$ psi.

Although the yield stress of the material is $\sigma_y = 30{,}000$ psi, elastic failure for this particular structure occurs at $\sigma_{cr} = 20{,}000$ psi.

Example 2: Statically indeterminate structure

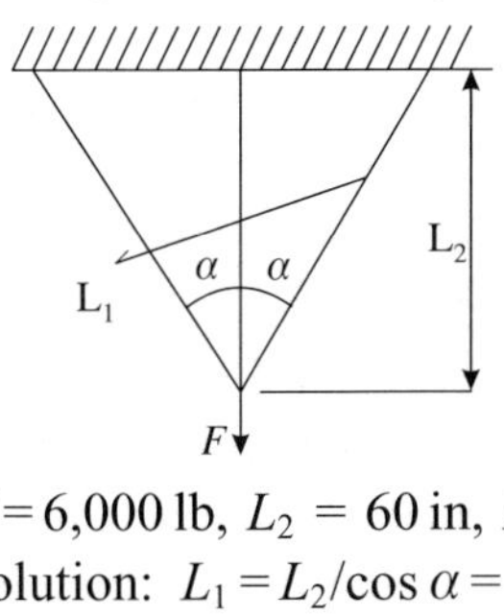

$F = 6{,}000$ lb, $L_2 = 60$ in, $E = 30\times10^6$ psi, $\sigma = 30{,}000$ psi, α=30°.
Solution: $L_1 = L_2/\cos\alpha = 69.3$ in. Elongation of the bar is $e_1 = N_1/ER_1$, $e_2 = N/ER_2$, $e_1 = e_2\cos\alpha$, $R_1 = A/L_1$, $R_2 = A/L_2 = A/L_1\cos\alpha$.
After substitution, $N_1L_1/EA = N_2\cos\alpha L_1\cos\alpha/EA$, $N_1 = N_2\cos^2\alpha$.
Static equation is $F = 2N_1\cos\alpha + N_2$,
$N_2 = F/(1 + 2\cos^2\alpha) = 6000/2.5 = 2{,}400$ lbs, $N_1 = 2{,}400 * 0.75 = 1{,}800$ lbs.

Optimal geometrical stiffness for $C_{st} = 1.7$ (tan 60°),
$R_o = \sqrt{N_2/EC_{st}} = \sqrt{2400/[30 \times 10^6 * 1.7]} = 0.007$ in.
Optimal cross-section of the bar, $A_o = R_o L_2 = 60 * 0.007 = 0.42$ in^2.
Corresponding stress, $\sigma_2 = N_2/A_o = 2400/0.42 = 5{,}714$ psi;
$R_{cr} = \sqrt{2400/30 \times 10^6 * 4} = 0.0045$ in, $A_{cr} = 60 * 0.0045 = 0.27$ in^2. Critical stress is $\sigma_{cr} = N_2/A_{cr} = 8{,}889$ psi.

Although the material allows stress $\sigma_y = 30{,}000$ psi the real structure has limiting stress $\sigma_{cr} = 8{,}889$ psi. The factor of safety in comparison with the yield stress of the material is $n = \sigma_y/\sigma_{max} = 30000/5714 = 5.25$. This is not the most efficient structure in terms of utilizing the elastic energy of the material. But the engineer should know the real limit.

The correctness of the theory can be tested and experimentally supported. To that end the deformation of the actual structure is measured. This allows more accurate calculation of the internal force in a structure or in part of a structure. The optimization of the dimensions of a structure depends on knowledge of the elastic force. We can evaluate a complex structure when the external forces are unknown. To that end we apply an experimental tensile force N to the bar, measure its deformation e and calculate the actual geometrical stiffness of the bar, $R_a = N/Ee$. If the deformation resulting from the actual forces is e_a, then the elastic force in this structure is $N_a = e_a E R_a$. The optimal geometrical stiffness required by the force N_a is $R_o = \sqrt{N_a/EC_s}$. The cross-sectional dimensions of the structure can be corrected accordingly with the ratio $A_a/A_o = R_a/R_o$, or $A_o = A_a R_o/R_a$.

Engineering calculations are always approximate. However the results will be adequate if the calculations are based on a proven reliable physical foundation. The actual individual limit for a bar can be calculated with the equation $N_{cr} = 4ER^2$ and then tested, and the prediction is supported experimentally. Therefore the system consisting of the basic and the derivative equation of the non-linear theory of elasticity is necessary and sufficient for reliable structural analysis and design.

3. Torsion

A twisting moment M applied to a bar causes shear deformation and shear stress in the cross-sections of the bar. The cross-sections of a round bar that are plane before twisting remain plane after twisting. The angle of torque of a cross-section is denoted by θ. For a round bar, a straight line along the bar becomes a helix in torsion. The elongation along this helix is e. The angle of twist along the bar is denoted by γ. The external twisting moment M is equal to the sum of the elastic moments in a cross-section of the bar. The elastic moment of resistance M_τ is equal to the sum of the products of the shearing stresses acting in a cross-section on the corresponding distances from the axis.

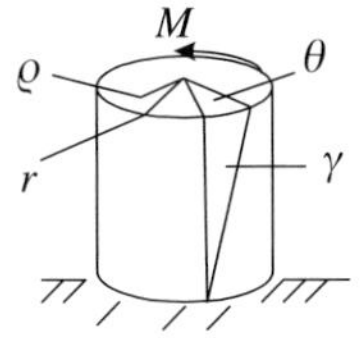

For example, for a shaft transmitting a power N (hp) and rotating at n (r/min) the twisting moment is $M = 63{,}030 * N/n$ (lb-in). For a round bar the total angular deformation is $\theta = M_\tau/GR_\tau$, where M_τ is the shear moment, G is the shear modulus of elasticity, and R_τ is the geometrical stiffness of the bar in torsion. For simplicity some indexes will be omitted. $R = I_p/L$, where I_p is the polar moment of inertia. For a round cross-section, $I_p = \pi d^4/32$. An important geometrical characteristic in torsion is the polar section modulus S_p. For a round bar, usually called a shaft, the polar section modulus is $S_p = \pi d^3/16$. For a tube, $I_p = \pi D^4(1 - c^4)/32$, $S_p = \pi D^3(1 - c^4)/16$, where $c = d/D$ is the ratio of inside diameter to outside diameter of the tube.

The formula for the rate of deformation is written accordingly, $d\theta/dR = -M/GR^2$. The equation of elastic stability in torsion is $M/GR^2 = C_s$. Optimal geometrical stiffness can be determined with the equation $R_o = \sqrt{M/GC_s}$. Then, the optimal polar moment of inertia is $I_{p,o} = R_o L$.

In order to find the elastic limit for a structure we have to compare the individual limit of elasticity depending on the geometry of a structure, which is obtained mathematically, with the limit of elasticity of the material, which is obtained experimentally. The actual limit of elasticity of a shaft is the lesser of these two values. The maximum shear stress in a cross-section should be less than the actual limit of elasticity, $\tau = M/S_p < \tau_{cr}$ or $\tau < \tau_y$.

The geometrical stiffness of a bar can be measured by applying to the bar an experimental moment M and measuring the total deformation θ, $R_a = M/G\theta$. For a length L the actual moment of inertia can be calculated as $I_p = R_a L$. For the given M the optimal geometrical stiffness of a shaft can be calculated, selecting the coefficient of elastic stability, for example $C_s = 1$, then $R_o = \sqrt{M/G}$. The stress for this shaft is $\tau_o = M/S_p$. The critical stress due to the geometry corresponds to $R_{cr} = \sqrt{M/GC_s}$, where $C_s = 4$, and the limit in this case is $\tau_{cr} = 2\tau_o$. The stress in the shaft should also be less than the limit of elasticity of the material, $\tau_o < \tau_y$.

The calculation of the optimal dimensions and stress in a shaft requires the description of the polar moment of inertia and the polar cross-section modulus. For all shapes of cross-section of a bar the stress for torsion and the total deformation of the bar are obtained with formulae similar to those for a round bar. However, "it has been found from many experiments that in non-circular sections the shearing unit stresses are not proportional to their distances from the axis. Thus in a rectangular bar there is no shearing stress at the corners of the sections and the stress at the middle of the wide side is greater than at the middle of the narrow side. In an elliptical bar the shearing stress is greater along the flat side than at the round side." (Handbook of Engineering Fundamentals, Edited by Mott Souders and Ovid W. Eshbach).

These experimental findings raise doubt about the assumption that the shearing unit stresses are proportional to the distances from the axis for any type of structure, i.e. round or non-round. The assumption here is that shearing stresses distributed in a cross-section have the same value, $\tau = M/S_p$, that is independent from the distance of a point from the axis. It is also assumped that shear stresses in torsion are tangential to circular lines drawn through the point in the cross-section and having their center on the axis of the bar. The direction of stresses is opposite to the direction of the external twisting moment. The integral sum of the distributed elastic moments is equal to the external twisting moment, $M = \int \tau \, dA\rho$.

The mathematical basis of this proposed theory of stress distribution is as follows. Let us consider angular deformation in a round bar. The correlation between the angular deformation γ along the bar and the corresponding twisting angle of cross-section at a distance ρ is $\tan \gamma_\rho = \rho\theta/L$. For a small deformation, $\gamma_\rho = \rho\theta/L$. At a distance r,

$\gamma_r = r\theta/L$. Then the ratio of the angular deformations after algebraic transformation $\gamma_r/\gamma_\rho = r/\rho$. The angular deformation, on the other hand, is proportional to the unit elastic moment, $m = G\gamma$. At the distance ρ from the axis, $m_\rho = \tau_\rho \mathrm{d}A\rho$, where τ_ρ is the unit stress tangential to the circular line drawn through the point with radius ρ, and $\tau\,\mathrm{d}A$ is the elemental tangential force. The unit elastic moment at the distance of radius of cross-section r is $m_r = \tau_r \mathrm{d}Ar$. The ratio of the elemental moments in cross-section at the different distances from the axis is $m_r/m_\rho = \tau_r r/\tau_\rho \rho$, or $\tau_r r/\tau_\rho \rho = G\gamma_r/G\gamma_\rho = r/\rho$. The equality $\tau_r r/\tau_\rho \rho = r/\rho$ is true only if $\tau_\mathrm{r} = \tau_\rho$. Then the conclusion is that the unit tangential stresses distributed in the cross-section of a round bar have the same value $\tau = M/S_\mathrm{p}$. Shearing stresses of equal magnitude are acting in the direction opposite to the twisting moment, tangential to the circular lines drawn from the axis. This theory explains the distribution of stress and deformation in non-round bars and explains the experimental findings described in the literature for non-round shapes.

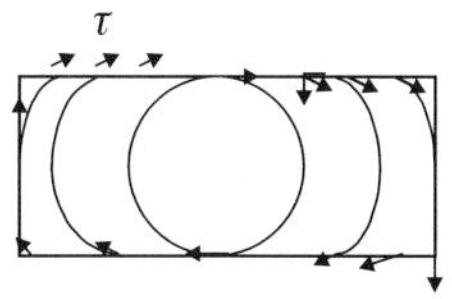

The drawing above shows the distribution of the stresses in a rectangular bar subjected to torsion. It explains the practical absence of stresses in the corners and the concentration of stresses at the surfaces of the long sides of a rectangle. The theoretical determination of stresses and deformations in bars with a non-round cross-section is presently complex and often inaccurate. The same applies to the determination of the effect of the polar moment of inertia on deformation. However, the experimental method of determinating the geometrical stiffness and then calculating the actual polar moment of inertia will provide more accurate knowledge of prismatic bars in torsion. We can calculate the geometrical stiffness of a bar by measuring the angular deformation for any type of bar, $R_\theta = M/G\theta$. The actual polar moment of inertia of the cross-section and the polar cross-section modulus can be obtained with the following equations: $KI_\mathrm{p} = R_\theta L$ or $R_\mathrm{e} = M/Ge$, where e is the elongation along the helix line, and $KS_\mathrm{p} = R_\mathrm{e}L$. Also, the polar moment of inertia and the section modulus can be calculated with formulae from the tables found in the handbooks. The coefficient K corrects the theoretical characteristic of a cross-section: it accounts for the specifics of the real cross-section. This coefficient can be found by experimentally obtaining the real KI_p $(=R_\theta L)$ calculating the theoretical I_p, from $K=R_\theta L/I_p$.

Example of shaft calculation: A steel shaft has two parts with lengths $L_1=96$ in and $L_2=60$ in. We need to determine the necessary optimal diameters of these parts, D_1 and D_2. The parts are subjected to torque moments $M_1=17{,}000$ in-lb and $M_2=-5{,}000$ in-lb; $E_{st}=30\times10^6$ psi, $\tau_y=25{,}000$ psi; $C_s=1$.

Then optimal stiffness of the shaft refers to an equivalent shaft of total length $L=L_1+L_2$, subjected to a moment M equal to the sum of the moments, $M=\sum M_i$. $\theta=M/GR$; $R_o=\sqrt{M/GC_s}=\sqrt{12{,}000/12\times10^6*1}=$ 0.032 in^3; $R_{cr}=0.016$ in^3; $I_{p,cr}=0.016*156=2.5$ in^4; $I_p=0.1D^4$; $D_{eq,cr}=$ 2.24 in; $S_{p,cr}=2.25$ in^3; $\tau_{cr}=12{,}000/2.25=5{,}333$ psi. Although the material allows stress $\tau_y=25{,}000$ psi the real limit for this structure is $\tau_{cr}=5{,}333$ psi. The optimal resistance of each part is equal to the resistance of the equivalent shaft. The common geometrical stiffness is $R_o=R_1=R_2$. This solution eliminates the relative shear of the parts to each other. The total deformation of the shaft is equal to sum of the angular deformations of the parts and equal to $\theta=12{,}000/(12\times10^6*0.032)=0.031$. The optimal moment of inertia for each part can be calculated with the equation $R=I_p/L$: $I_1=R_oL_1=0.032*96=3.1$ in^4, $I_2=0.032*60=1.9$ in^4. The diameter of the parts can be calculated accordingly, $I_p=0.1D^4$. The optimal diameters of the parts are $D_{1,o}=2.4$ in, $D_{2,o}=2.1$ in. The section modulus of the parts is $S_p=0.2D^3$. Calculation yields $S_1=2.7$ in^3, $S_2=1.9$ in^3. The torsion stresses are $\tau=M/S_p$, $\tau_1=12{,}000/2.7=4{,}444$ psi, $\tau_2=5{,}000/1.9=2{,}632$ psi. Each part has a stress that is both less than the yield stress of the material and less than the critical stress for this shaft.

3.1 Recapture

For a bar subjected to torsion the unit tangential stresses in a uniform cross-section have the same value $\tau = M/S_p$ that does not depend on the distance of the point from the axis of rotation.

The unit elastic moment, on the other hand, does depend on the distance of the point from the axis, $m_\rho = \tau(\Delta A)\rho$. The shear stresses are acting in the direction opposite to the twisting moment, tangential to circular lines drawn around the axis and through the points.

This mathematical conclusion necessarily follows from the equation connecting the angular displacement along a bar and the angular deformation of its cross-section. Although this equation is known the conclusion is presented for the first time.

This theory of stress distribution in a bar in torsion explains the experimental findings of stress and deformation in non-round bars.

Our method of structure optimization allows us to obtain more adequate knowledge of the geometrical stiffness of bars having different shapes, more correct calculation of the individual allowable stresses with the equation of elastic stability, and optimal dimensions of a bar subjected to torsion.

4. Bending

A bar subjected to bending is called a beam. The external forces causing bending are directed perpendicular to the axis of the beam. The bending deformation is characterized by elongation of one side of the beam and compression of the opposite side. The corresponding stresses are normal to the cross-section and form couples of the bending–resistance moments. It is an accepted assumption that a beam cross-section that is plane and perpendicular to the beam axis before deformation remains plane and perpendicular to the axis after deformation. The straight axis of the beam, however, after deformation becomes a curve and the plane of a cross-section is at an angle to its initial position. Shear deformations cause shear stresses in the cross-sections. The distribution of shear stresses in a cross-section is uneven. The maximum shear stress is in the middle of the cross-section, while at the surface of the beam the shear stresses are zero. The shear stress in a cross-section can be calculated with the formula $\tau_{zy} = Q_y S / I_x b$, where S is the static moment of the cross-section and b is the width of the cross-section. The maximum tangential stress is at the neutral line of the cross-section.

Normal stresses in a cross-section from the side of elongation are considered positive (+). Normal stresses from the side of contraction are considered negative (–). The elastic moments from these stresses with respect to the central axis are equal and have opposite signs/directions. The normal elastic moments in a cross-section have a linear distribution. The maximum moment is at the greatest distance from the central axis; the moment is zero at the axis. The coordinates are drawn through the center of gravity of the cross-section. The normal stress at any point of the cross-section can be calculated with the formula $\sigma = M/S$, where S is the section modulus. The curvature of the axis of the beam is $1/\rho = M/EI$.

Bending can be simple if the bending moments are acting only in one of the main planes (xz or yx). If the bending moments act in two planes the bending is considered complex. The stress and deformation in this case can be obtained as the sum of the corresponding stresses and deformations in two planes.

We can find the bending deformation using the equation for the elastic line, $EI\mathrm{d}^2y/\mathrm{d}x^2 = M$, where y is the vertical displacement of the beam at a distance x from the origin of the coordinates. The angular deformation of a cross-section is proportional to the normal stress in the cross-section, $\varepsilon = \sigma/E = M/EI$, where $\varepsilon = \mathrm{d}\theta/\mathrm{d}x$. Thus, $\theta = ML/EI$ or total angular deformation of a beam is $\theta = M/ER$, where the more correct expression of geometrical stiffness is $R = KI/L$.

The system of functions for the calculation of optimal beam design consists of the basic equation $\theta = M/ER$ and the derivative equation $\mathrm{d}\theta/\mathrm{d}R = -M/ER^2$. The elastic moment M in these equations is the arithmetic mean value of the moments distributed in the structure. If the external moment $M_{\max}$ is calculated with the equations of static equilibrium, then the elastic moment is $M = M_{\max}/2$. The optimal geometrical stiffness of a beam can be found with the equation of elastic stability, $M/ER^2 = C_s$. The coefficient of elastic stability C_s is selected with the knowledge of the elastic relations as described in the non-linear theory with the system of two non-linear equations. A graph of θ vs. R helps to visualize the proper selection. The optimal geometrical stiffness is $R_o = \sqrt{M/EC_s}$. The geometrical stiffness is $R = KI/L$, where

K is the experimental coefficient of beam design specifics. For $K=1$ the optimal moment of inertia of the cross-section is $I_o = R_o L$. The geometrical stiffness of a real structure can be obtained experimentally by measuring the deformation θ from an experimental force/moment, $R_a = M/E\theta$. The actual geometrical stiffness can then be compared with the calculated optimal geometrical stiffness in order to find the correct dimensions for the structure, $R_o/R_a = I_o/I_a$. The structure-specific critical stress that corresponds to the critical value of geometrical stiffness can be calculated. The coefficient of elastic stability corresponding to the limit is approximately $C_s = 4$ (tan 76°); thus $R_{cr} = \sqrt{M/EC_s} = 0.5\sqrt{M/E}$, $I_{cr} = R_{cr}L$, and S_{cr} corresponds to I_{cr}, then $\sigma_{cr} = M/S_{cr}$. The stress in a structure of optimal dimensions should be less than the critical stress required by the condition of elastic stability and less than the proportional limit of the material σ_y, that is, $\sigma_o = M/S_o < \sigma_{cr}$ and $\sigma_o < \sigma_y$.

Example: For a simple steel beam with $E = 30 \times 10^6$ psi, $\sigma_y = 60{,}000$ psi, we have to calculate the optimal moment of inertia of the cross-section, I_o.

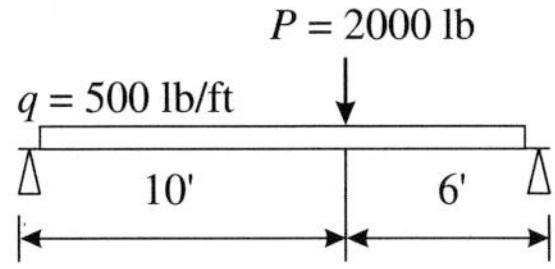

The theoretical maximum bending moment is $M_{max} = M_1 + M_2$, where $M_1 = 2000(16-10) * 10/16 = 7{,}500$ lb-ft, $M_2 = \frac{1}{8}qL^2 = 16{,}000$ lb-ft; $M_{max} = 23{,}500$ lb-ft $= 282{,}000$ lb-in; $M = M_{max}/2 = 141{,}000$ lb-in. Selecting the coefficient of elastic stability $C_s = 1$, we have $R_o = \sqrt{M/EC_{st}} = \sqrt{141 \times 10^3/30 \times 10^6} = 0.07$ in^3; $I_o = R_o L$ in$^4 = 0.07 * 16 * 12 = 13.4$ in^4. The table of American Standard Beams shows that close to the calculated value we can select $I = 15$ in^4 and $S = 6$ in^3. The stress in a structure of optimal stiffness is $\sigma_o = M/S = 141{,}000/6 = 23{,}500$ psi.

The structure-specific critical stress corresponds to the critical value of geometrical stiffness, $R_{cr} = 0.5\sqrt{M/E} = 0.034$ in^3. Selection from the standards gives $I = 6.5$ in^4, $S = 3.3$ in^3; $\sigma_{cr} = 141{,}000/3.3 = 42{,}727$ psi. The bending stress in a beam of optimal geometrical stiffness is less than the structure-specific critical stress and less than the yield stress of the material. The safety factor with respect to yield stress is

$n = 60{,}000/23{,}500 = 2.6$. The safety factor with respect to the structure-specific critical stress is $n = 42{,}727/23{,}500 = 1.8$. For a beam of optimal dimensions the angular deformation is $\theta = M/ER = 141 \times 10^3/(30 \times 10^6 * 0.07) = 0.07$.

The maximum deflection $Y = Y_1 + Y_2$. For calculating of the maximum vertical deflection in the middle of the beam, $x = L/2$, we use the formulae that are obtained with the equation for the elastic line. For this example,

$$Y_1 = \frac{P}{6EI}\left[\frac{(L-a)(\frac{1}{2}L)^3}{L} - (\tfrac{1}{2}L - a)^3 + \tfrac{1}{2}(L-a)^3 - \tfrac{1}{2}(L-a)\mathrm{L}^2\right].$$

Calculation yields $Y_1 = 0.0046$ in. For the distributed load q the deflection is $Y_2 = 5qL^4/384EI = 1.83$ in. The total deflection of the beam is $Y = 1.84$ in.

The equation of angular deformation and the equation for the elastic line can be used for calculating deformations and deflections. Note, however, that these equations are not suitable for calculating the optimal dimensions of a structure based on the condition of elastic stability. The difference between the equation for the elastic line in LTE and the equation of deformation in NLTE is that the former refers to a physical body, while the latter describes both an individual structure and a set of similar structures.

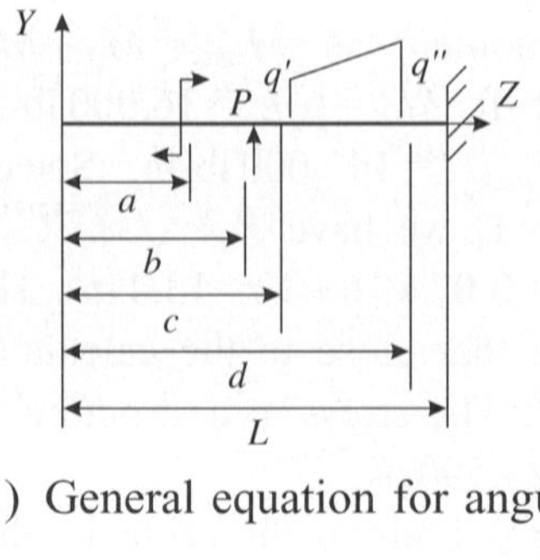

(1) General equation for angular deformations:

$$\theta = \frac{1}{EI}\Big[C + M(z-a) + \tfrac{1}{2}P(z-b)^2 + \tfrac{1}{6}q'(z-c)^3 + \tfrac{1}{24}k(z-c)^4 - \tfrac{1}{6}q''(z-d)^3 - \tfrac{1}{24}k(z-d)^4\Big].$$

(2) Equation for the elastic line:

$$Y = \frac{1}{EI}\Big[D + Cx + \tfrac{1}{2}M(z-a)^2 + \tfrac{1}{6}P(z-b)^3 + \tfrac{1}{24}q'(z-c)^4 + \tfrac{1}{120}k(z-c)^5 - \tfrac{1}{24}q''(z-d)^4 - \tfrac{1}{120}k(z-d)^5\Big].$$

In these equations: $C = EI\theta_o$, $D = EIY_o$, $k = (q'' - q')/(d - c)$. The coefficient k accounts for a linear change of the distributed load.

Although the equation of angular deformation for θ is a partial differential equation of the equation for the elastic line Y, both these equations can be tested experimentally, for they both describe physical changes in a structure. This is different from the non-linear equation of deformation and the partial derivative equation of elastic stability. The equation of deformation in the non-linear theory is used for the mathematical analysis of elastic behavior of similar structures. That equation is a mathematical curve and its derivative is also a mathematical curve. The end result of this derivative is a mathematical property that can be proved or disproved by mathematical means. On the contrary, the physical property θ can be measured as well as the deflection Y.

4.1 Calculation of deflections using the unit load method

A convenient way to calculate displacements is the unit load method. The method is based on Castigliano's theorem. If a structure is in static equilibrium then displacements of the points of application of the forces/moments in the direction of the forces are equal to the corresponding partial derivatives of potential energy of deformation: $\delta = \partial U/\partial P$ or $\theta = \partial U/\partial M$.

The unit load method allows one to find the displacement at any point of the elastic line from the actual elastic forces and applying the unit force at the point in which we determine the displacement:

$$\delta = \frac{\int M_x M_{x1}\,\mathrm{d}l}{EI_x} + \frac{\int M_y M_{y1}\,\mathrm{d}l}{EI_y} + \frac{\int M_T M_{T1}\,\mathrm{d}l}{GI_p} + \frac{\int N N_1\,\mathrm{d}l}{EA} + k_x\frac{\int Q_x Q_{x1}\,\mathrm{d}l}{GA} + k_y\frac{\int Q_y Q_{y1}\,\mathrm{dl}}{GA},$$

where the index 1 indicates the unit force, the coefficient k depends on the shape of the cross-section, and $\mathrm{d}l$ is the element of length.

This formula can serve only for determining the displacements in a structure of known geometrical parameters. It cannot be used for optimizing a structure without resort to the equation $\theta R = \sum M_i M_1/E$. Then $R_o = \sqrt{\sum M_i M_1/EC_s}$.

In the Castigliano method the assumption is made that the distribution of elastic forces follows the distribution of external forces. This assumption contradicts the principle of continuity of deformation in a body. The deformation in a body is distributed evenly with the exclusion of the stress-raising places.

5. Combined stresses

A combination of bending with tension or compression occurs, for example, when a force is directed parallel to, but is not coincident with, the axis of the bar. It can also be a combination of forces acting perpendicular to the axis with a force acting along the axis.

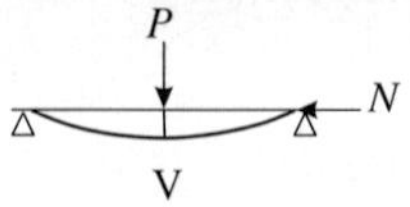

In the general case such loading gives the moments $M_y = Px$, $M_x = Py$ and the normal centric force N.

In a beam subjected to bending and a compressive force acting along the axis this force causes not only compression but also an additional deflection of the beam. The maximum normal stress in a cross-section of the beam is equal to the sum of the stresses, $\sigma_{max} = N/A + M_x/S_x + NV/S_x$. The total deformation of the beam is equal to the sum of the deformations from these forces. This sum can be presented as a the deformation of an equivalent beam with maximum stress equal to the actual σ_{max} and the corresponding equivalent moment, $M_{eq} = \sigma_{max} S_x$. Then the basic equation of deformation is $\theta = M_{eq}/ER$, the equation of elastic stability is $M_{eq}/ER^2 = C_s$, the optimal geometrical stiffness is $R_o = \sqrt{M_{eq}/EC_s}$, $I_o = R_o L$, $\sigma_{max} = M_{eq}/S_o < \sigma_{cr}$, and $\sigma_{max} < \sigma_y$.

Beams and columns, in which the axial forces may produce the bending moment, are presently checked for elastic stability with the empirical formulae. These formulae are based on Euler's equation for the critical force for a column, $P_{cr} = k\pi^2 EI/L^2$, where k is an empirical coefficient depending on the type of supports of a column and other design-specific features.

In the non-linear theory of elasticity for calculations on structures that can suddenly collapse from sidewise bending, we will use our basic equation of elastic stability for calculating the optimal geometrical stiffness of a structure.

Example: A steel column, $\sigma = 60{,}000$ psi, supports a total load $P = 200{,}000$ lb. The length of the column is $L = 15' = 180$ in.

The optimal dimensions of the column can be found with the equation of elastic stability for tension/compression, $R_o = \sqrt{P/E} = \sqrt{200,000/30\times10^6} = 0.082$ in $= A_o/L$, giving $A_o = 14.8$ in^2. The tensile stress in the column is $\sigma = P/A = 13{,}513$ psi. The critical stress corresponds to the critical value of geometrical stiffness, $R_{cr} = \sqrt{P/EC_s}$, so with $C_s = 4$ we have $R_{cr} = 0.041$ in; then $A_{cr} = R_{cr}L = 7.4$ in^2; the structure-specific critical stress is $\sigma_{cr} = 27{,}026$ psi. Selecting a wide-flange section from the table of structural sizes, 8×8, weight 48 lb/ft gives $A = 14.11$.

If we assume that bending of the column may occur, then we may use the equation for the elastic line for a beam with one end fixed, $Y_{max} = ML^2/2EI$. The bending moment with eccentricity of load equal to the maximum deflection is $M = PY_{max}$, and the equation becomes $Y_{max} = PY_{max}L^2/2EI$ or $1 = PL^2/2EI$. This formula is similar to Euler's column formula, $P_{cr} = 2EI/L^2$. The selection of the moment of inertia depends on the force P and does not depend on the possible Y_{max}: $I = PL^2/2E = (2\times10^5 * 180^2)/(2 * 30\times10^6) = 108$ in^4. Selection from the standards gives the wide-flange section 8×8 and corresponding section modulus $S = 27.4$ in^3. The maximum stress in this column is $\sigma = M/S$. The maximum eccentricity of the load for the selected section is $e = b/2 = 4$ in, $M = Pe = 800{,}000$ lb-in. The maximum stress is $\sigma_{max} = 800{,}000/27.4 = 29{,}197$ psi. The critical stress based on the condition of elastic stability is $\sigma_{cr} = 27{,}026$ psi, less than the possible σ_{max} in case of bending. Therefore we may select the same 8×8 size but with weight $W = 58$ lb/ft, $A = 17.06$ in^2, $S = 52$ in^3. Then the maximum stress in the column, $\sigma = 15{,}384$ psi, is less than σ_{cr}.

The difference between Euler's experimental column formulae and the equation of elastic stability is that the equation of elastic stability is a general equation that always produces reliable results. Euler's experimental formulae are not as accurate and reliable.

6. Continuous beam

A continuous beam, i.e. a beam that has more than two supports, is statically indeterminate. The reactions in the supports of a continuous beam cannot be obtained with the equations of static equilibrium only. For the calculation of the reactions in the supports each section is considered as an independent beam. The action of the adjacent part is substituted by a moment in the support between the sections. The equation for determining these moments is known as the theorem of three moments.

A beam that has one fixed end and n roller supports is n-fold statically indeterminate. If both ends of the beam are fixed, then the degree of indeterminacy is equal to the number of supports. The moment in a roller support for a beam with a console is equal to the moment from the load on the console. We calculate the necessary geometrical stiffness of the beam with the equation of elastic stability, $R = \sqrt{M/EC_s}$. The stress in the beam should be less than the beam-specific and material limits, $\sigma_{max} < \sigma_{cr}$ and $\sigma_{max} < \sigma_y$.

Example:

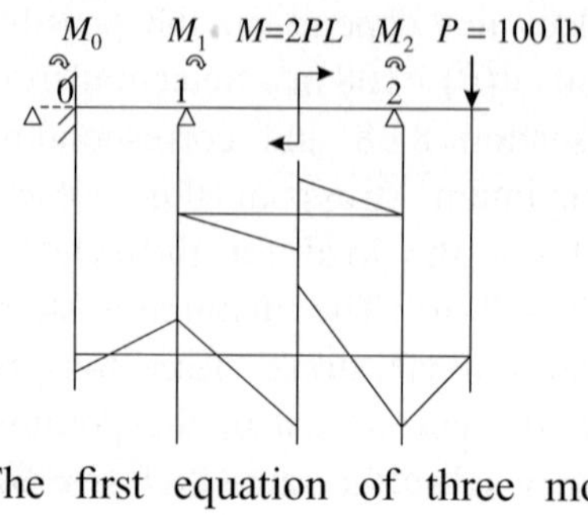

The first equation of three moments includes an artificial section ($M = 0$) and section 1: here $2M_0L + M_1L = 0$. This sum is equal to 0 because the section has no active load. The second equation of three moments relates to sections 1 and 2: $M_0L_1 + 2M_1(L_1 + L_2) + M_2L_2 = -6(\Omega_1 a_1/L_1 + \Omega_2 a_2/L_2)$, where Ω_i is the area of the diagram of moments for section i and a is the coordinate of center of gravity of the area.

Here, $M_2 = -PL_3$; $L_1 = L_2 = 50$ in; $L_3 = 25$ in, $a' = \frac{2}{3}L$, $a'' = \frac{1}{3}L$. After some substitutions the equations become $\frac{7}{2}M_1 + 2500 = -6(PL * L/4) * (a' + a'')/L = \frac{1}{4}PL - 2{,}500$; $M_1 = 5{,}000$ lb-in, $M_{max} = 5{,}000$ lb-in; the critical geometrical stiffness is $R_{cr} = \sqrt{M/EC_s} = 0.0065\ \text{in}^3$; $I_{cr} = 0.325\ \text{in}^4$; $d_{cr} = 1.6$ in; $S_{cr} = 0.41\ \text{in}^3$; $\sigma_{cr} = 12{,}145$ psi; $d_o = 3.2$ in; $\sigma_{cr} < \sigma_y$.

The actual geometrical stiffness of the sections of a continuous beam and the distributed elastic forces in these sections can be obtained by measuring the deflections of the individual sections. However, the analysis of deformation and geometrical stiffness cannot be performed within the linear theory of elasticity. Such analysis can be done only within the system of the equations of the non-linear theory of elasticity.

7. Stability of thin shells

A shell is a structure that can be described by curved surfaces of relatively small thickness. Several sophisticated theories of thin shells describe mathematical procedures for obtaining stresses. It is difficult to validate their practical importance. A common explanation is that it is expensive to build a real shell in order to check stresses. Also, no criterion of similarity existst for evaluating the experimental data.

In the non-linear theory of elasticity we do have a method that provides similarity of the structures. This method also allows us to calculate the optimal dimensions of the structures in the range of elastic stability. The criterion that provides optimization and similarity of the structures describes the rate of change of deformation. For shells, just as for the structures dealt with in the preceding text, we need an equation of deformation and a derivative equation that describes the relation between internal force and geometrical stiffness within the limit of elastic stability. At present, empirical buckling formulae are used to determine the critical load for a shell. In the non-linear theory we have scientific formulae which allow us to calculate the necessary thickness for a given load. The geometrical parameters of a structure can be changed in order to obtain optimal geometrical stiffness. Formulae for the stresses in a shell are mathematically well evaluated.

There are two main shell theories: membrane theory and bending theory. If a shell is subjected only to distributed pressure (no concentrated loads and moments applied to the shell), then we deal only with normal stress in each section of a shell element. If bending moments may rise in a shell, these moments should be considered in calculation of the dimensions.

The total elongation, for example, in the meridian direction L of a shell is $e=M/ER$, where $R=KS/L$ and S is the section modulus of the respective cross-section. The equation of elastic stability is accordingly $M/ER^2=C_s$. The geometrical stiffness of a model of a shell can be measured, $R=M/Ee$. Comparison of the actual geometrical stiffness of the model and the geometrical stiffness calculated with the equation of elastic stability allows us to correct the geometrical descriptions of R and S.

The method of optimization based on the non-linear theory of elasticity is a general and reliable method for any type of a structure. The coefficient of elastic stability provides similarity of results. The method allows one to connect description and experimental data. Experiments without the sound theory are insufficient as a source of knowledge. As well, theory without experimental support is insufficient for reliable engineering practice.

In a general case the equations connecting internal forces or moments and deformation can be obtained with a generalized form of Hooke's law.

7.1 Calculation for symmetrical thin shells

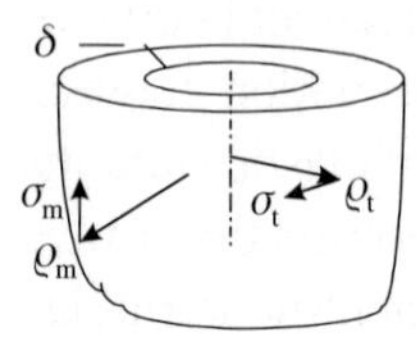

A symmetrical thin shell that is subjected to internal or external pressure has only normal stresses distributed in the material of the wall. These stresses can be obtained with membrane theory. Stresses in the meridian section of a shell and in the cross-section of the shell are connected by Laplace's formula, $\sigma_m/\rho_m + \sigma_t/\rho_t = p/\delta$. A second equation for determining stresses is obtained from the condition of equilibrium of stresses and forces/pressure in the part of the shell under consideration: $\sigma_m 2pr\delta\alpha = \sum P_x$, where $\sum P_x$ is the sum of the projections of the external forces acting on the part. For calculating the optimal geometrical stiffness and the optimal thickness of a shell we need the equation of deformation and the equation of elastic stability. The equation of deformation can be obtained by substituting $\sigma = E\varepsilon$ and $\varepsilon = eL$. The equation then becomes $Ee2\pi r\delta \cos\alpha/L = \sum P$. The geometrical stiffness of the shell is $R = 2\pi r\delta/L$. The maximum elongation of a meridian line is $e_{max} = \sum P_{max}/ER_{min}$. The equation of elastic stability is $\sum P/ER^2 = C_s$.

The optimal geometrical stiffness for the selected coefficient of elastic stability $C_s = 1$ is $R_o = \sqrt{\sum P/E}$; the optimal thickness of the shell is $\delta_o = R_o L/2\pi r$; the meridian stress is $\sigma_m = \sum P/2\pi r\delta$; and $\sigma_{cr} = 2\sigma_o$.

The tangential stress can be calculated as $\sigma_t = (p/\delta - \sigma_m/\rho)\,r$, and should be lower than the critical stress, $\sigma_t < \sigma_{cr}$.

The method of optimization is a general method that provides similarity for a series of similar structures. We can obtain the actual geometrical stiffness of a structure, $R_a = F/Ee$, by measuring the total deformation on a prototype of the structure in the meridian or/and in the cross-sectional direction. Comparison the actual geometrical stiffness with the calculated optimal value, $R_o = \sqrt{F/EC_s}$, allows us to change the geometrical parameters so as to obtain the optimal design. The non-destructive method in addition allows us to study the effect of different geometrical parameters on the geometrical stiffness of the structure.

Structures with different shapes can have the same geometrical stiffness. The formulae for actual geometrical stiffness and optimal geometrical stiffness do not consider a particular geometry. The designer is free to select the geometry of the structure. The description of the geometrical stiffness for a particular structure allows one to change the parameters, such as length, cross-sectional dimensions, radius of curvature, shell thickness, etc., in order to achieve the optimal geometrical stiffness.

8. Elastic stability of plates

A plate is a structure with two plane surfaces and a thickness that is much less than the two other dimensions. At present, plates are divided in thin plates or membranes and thick plates. The imaginary surface that divides the plate in two halves of equal thickness is called the middle surface. The assumptions in the theory of thin plates are the following:

(1) Any straight lines that are perpendicular to the middle surface before deformation remain perpendicular to the middle surface after deformation.

(2) Stresses that are perpendicular to the middle surface can be neglected.

For a thick plate, the normal and tangential stresses can be obtained with Hooke's law: $\sigma_x = E(\varepsilon_x + \mu\varepsilon_y)/(1-\mu^2)$; $\sigma_y = E(\varepsilon_y + \mu\varepsilon_x)/(1-\mu^2)$; $\tau = E\gamma/2(1+\mu)$. The maximum stresses according to the linear theory of elasticity will be on the surfaces of the plate: $\sigma_x = 6M_x/at^2$; $\sigma_y = 6M_y/bt^2$; $\tau = 6H/at^2$. According to Hooke's law the equations can be written in terms of the deformations: $\theta_x = M_x/ER$; $\theta_y = M_y/ER$; $\gamma = H/ER$. Correspondingly, the geometrical stiffnesses are: $R_x = 6a/t^2$; $R_y = 6b/t^2$; $R_\gamma = 6a/t$, where a and b are lengths of the sides for the rectangular plate, and t is its thickness.

For a round plate the formulae change accordingly: $R = 6r/t^2$, $\theta_r = M/ER$, $\theta = (1-\mu)M/ER$, $\gamma = H/ER_\gamma$; maximum stress is $\sigma = 6M/rt^2$.

Example 1: Rectangular plate with four edges simply supported. $a = 60''$; $b = 30''$; uniformly distributed load $w = 1000$ psi; material is steel with $E = 30 \times 10^6$ psi, $\sigma_y = 60{,}000$ psi.

The task is to find the necessary thickness of the plate.

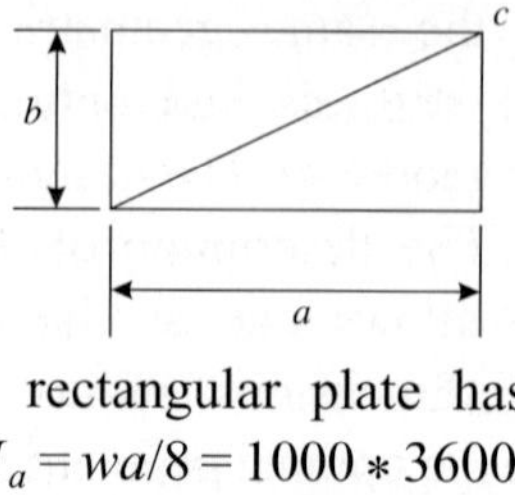

A rectangular plate has diagonal symmetry. Maximum moments are
$M_a = wa/8 = 1000 * 3600/8 = 4.5 \times 10^5$ in-lb;
$M_b = wb/8 = 1000 * 900/8 = 1.125 \times 10^5$ in-lb;

$M_{c,\max} = \sqrt{M_a + M_b} = \sqrt{(4.5 + 1.125)\times 10^5} = 4.6\times 10^5$ in-lb.
The geometrical stiffness required by $M_c = M_{\max}/2$ is $R_o = \sqrt{M_c/EC_s} = \sqrt{2.3\times 10^5/30\times 10^6 * 1} = 0.09$ in^3; $R = I/c = t^3/12$; $t = \sqrt[3]{12R} = 1.03$ in.
The maximum stress in the plate is in the diagonal section, $\sigma_c = M_c/S_c$. The assumption is that stress in a plate of uniform thickness is distributed evenly. The section modulus is $S_c = ct^2/6 = 11.2$ in^3, and $\sigma_c = 2.3\times 10^5/11.2 = 20{,}536$ psi. The critical stress from the condition of elastic stability is $\sigma_{cr} = 2\sigma = 41{,}071$ psi. The material with $\sigma_y = 60{,}000$ psi can thus be selected.

Example 2: Circular flat plate made of steel, with edge simply supported; radius $r = 60$ in; uniform load $p = 1{,}000$ psi, $E = 30\times 10^6$ psi, $\sigma_y = 60{,}000$ psi.

The task is to find the necessary thickness of the plate.

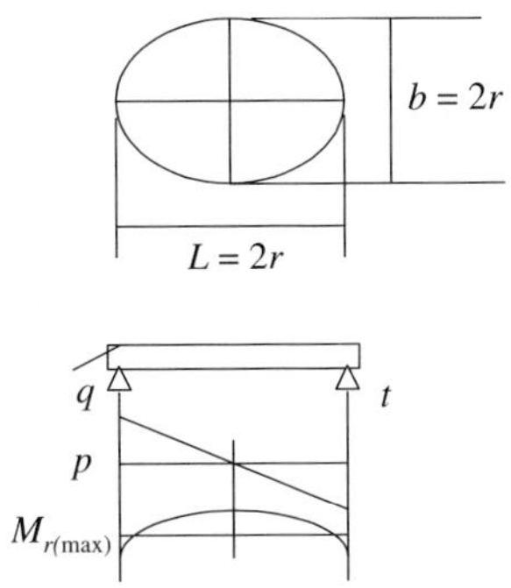

The uniformly distributed load causes internal radial and tangential moments correspondingly. The maximum moment $M_{\max} = \sqrt{M_r + M_\theta}$ is distributed per unit of length moment. From the condition of elastic equilibrium we can calculate the distributed elastic force q: $p\pi r^2 = q2\pi r$, $q = pr/2$. In our example, $q = 10^3 * 60/2 = 3\times 10^4$. The maximum elastic moment is found as follows: $M_r = M_\theta = qr/2 = pr^2/4 = 3\times 10^4 * 60/2 = 9\times 10^5$, $M_{\max} = 1.4M_r = 12.6\times 10^5$; $M = M_{\max}/2 = 6.3\times 10^5$. Note that the formula for the elastic moment in the linear theory is $M_r = M_\theta = (3+\mu)pr^2/16 = 0.21pr^2$ and we are on the safe side.

The optimal geometrical stiffness, $R = \sqrt{M/EC_s}$, for a coefficient of elastic stability $C_s = 0.58$ (tan 30°) is $R_o = \sqrt{6.3\times 10^5/30\times 10^6 * 0.58} = 0.19$.

The moment of inertia is $I_o = KR_o * 2r = 22.8\,\text{in}^4$, where K is an experimental coefficient of specifics depending on the boundary conditions. In this example we take $K = 1$ as no experiment was performed. From the moment of inertia, $I = bt^3/12$, we find the thickness as $t = \sqrt[3]{12I/2r} = 1.32$ in; the section modulus then is $S = 2rt^2/6 = 34.7\,\text{in}^3$. The stress in the plate is $\sigma = M/S$, thus $\sigma_{\max} = 6.3 \times 10^5/34.7 = 18{,}156$ psi. The maximum stress in such plate is less than the limit of elasticity of the material, $\sigma_{\max} < \sigma_y$. The angular deformation is $\theta_r = M/ER = 6.3 \times 10^5/(30 \times 10^6 * 0.19) = 0.11$.

The maximum deflection in the center of the plate is approximately $\upsilon_{\max} = \theta r/3 = 2.2$in, so $\upsilon_{\max} > t$. If we need $\upsilon_{\max} < t$, then we may increase t. For $t = 2$ in we have $I = 120 * 8/12 = 80\,\text{in}^4$, $R = I/b = 0.67$, $\theta = 0.03$, $\upsilon_{\max} = 0.62$ in. A test of the maximum deflection may prove the accuracy of the formulae.

The formula for calculating the maximum stress in the linear theory of elasticity is $\sigma_{\max} = 3pr^2(3 + v)/8t^2$. The calculation of thickness in this case is based on the maximum allowable stress $\sigma_{\max} = \sigma_y/n$. This yields $t = \sqrt{3 * 1000 * 3600 * 3.3/(8 * 30{,}000)} = 12.2$ in, which is 6 times more than suggested based on the condition of elastic stability. The maximum deflection in the current theory is given as $\upsilon_{\max} = 3p(1 - \mu)(5 + \mu)\,r^4/16Et^3$, where μ is Poisson's coefficient; for $\mu = 0.3$ and $t = 12.2$ in, we have $\upsilon_{\max} = 0.17$ in.

The current theory of plates operate with a characteristic D that designates the stiffness of a plate, $D = Et^3/12(1 - \mu^2)$. The geometrical stiffness of a plate in the non-linear theory is $R = KI/L = bt^3/12b = t^3/12$. Thus $D = ER$. If D accounts for the moment of inertia $I = bt^3/12$ then why is it that the formula for determining stress in the linear theory does not take account of the section modulus. In the non-linear theory $S = bt^2/6$, and $\sigma = M/S$.

Although the formulae for calculating plates were not tested, the method for plate calculation, i.e. establishing the connection between elastic moment and geometrical stiffness, is a general theoretical basis of design. Testing the formulae of geometrical stiffness and elastic moments includes measurement of the deformation from the experimental force/moment in a model. The formulae can be corrected with the coefficient of design specifics $K = \theta_{\exp}/\theta$. The formulae are acceptable if the result of calculation corresponds to the result of the measurement.

9. Dynamic stresses and the non-linear theory of elasticity

The engineer has the problem of predicting the stress at failure for parts of machines and structures that are subjected to dynamic forces. The calculation of strength and stability has the character of check-calculation. The non-linear theory of elasticity not only considers the mechanical characteristics of the material, but also the structure-specific critical stress that is generated by the geometry of the structure. Although no special research work has been done for structures with dynamic stresses, the basic formulae of non-linear theory have to be instituted in the current approach to such structures. The domain of possible stable relations is the interval of a tangent function that represents the rate of change of deformation $0 < d\delta_{dyn}/dR < 4$. In order to satisfy this condition the frequency of oscillation needs to be regulated accordingly.

Vibration stresses in bars and beams cause deflections, which are calculated with formulae that multiply the value of static deflection by a dynamic magnification factor. "The relation is given by $\delta_{dyn} = \delta_{st.} * 1/[1-(\omega/\omega_n)^2]$ where ω is the frequency of oscillation of the load and ω_n the natural frequency of oscillation of the bar." (Handbook of Engineering Fundamentals, edited by Mott Souders and Ovid W. Eshbach). The natural frequency is $\omega = \sqrt{EIg/L^3W}$, where I is the moment of inertia, g is the acceleration of gravity, L is the length of the bar, W is the weight of oscillating load, and E is the modulus of elasticity.

Thus, for finding the dynamic magnification factor f_d we need to establish the dimensions of the structure based on the static elastic stability approach in the non-linear theory of elasticity. The optimal geometrical stiffness is $R_o = \sqrt{M/EC_s}$. The coefficient of elastic stability C_s for dynamic stresses can be selected the same as for static stresses. The optimal moment of inertia is $I_o = R_oL$. Calculation of ω_n and the dynamic magnification factor then allows us to find the optimal cross-sectional dimensions accordingly, $\delta_d/\delta_{st} = R_d/R_{st} = f_d$, $R_d = R_{st}\, f_d$, $I_{d,o} = R_{d,o}L$. The maximum stress in a structure with dynamic stresses should be less than the limiting stress of the material for such cases and less than the structure-specific critical stress. The critical stress can be determined with the equation of elastic stability. The critical geometrical stiffness is R=$0.5\sqrt{M/E}$, where M is the mean value of bending moment increased by the dynamic magnification factor f_d. For the dynamic forces acting on the structure it is also necessary to evaluate the character of elastic forces acting in the structure. This evaluation includes a curve that describes changes of force/deformation in time, and also the rate of change of deformation in time. This examination may help to select a value of elastic force for the calculation of optimal geometrical stiffness with the equation of elastic stability. The critical section modulus can be calculated as $S_{cr} = R_{cr}L$, and the critical stress is $\sigma_{cr} = M/S_{cr}$. Confirmation of the geometrical stiffness of a bar in the range of elastic stability is especially important in the case of dynamic stresses. Otherwise in the range of elastic instability such stresses cause great deformations that destroy the structure.

The limiting characteristic of a material in the case of dynamic stresses is the fatigue endurance limit. The endurance stress corresponds to the stress that does not destroy a specimen during approximately 10^8 cycles. The maximum stress in the structure should be less than the critical stress and less than the endurance limit, $\sigma = M/S_o \leqslant \min(\sigma_{cr}, \sigma_d)$.

Design of structures subjected to dynamic forces with the equations of non-linear theory should provide more reliable structures. For both situations with dynamic stresses and situations with static stresses the theory has a good experimental method for confirming its results.

10. Impact stresses

Deformation and stress in a bar from a force of impact can be obtained with the formulae for the static force by multiplying with an experimental dynamic coefficient k_{dyn}. For simple cases this coefficient can be calculated. This equivalent static load method neglects the effect of wave propagation. The optimal geometrical stiffness of a bar can be found with the equation of elastic stability for equivalent load.

10.1 Tension impact on a bar

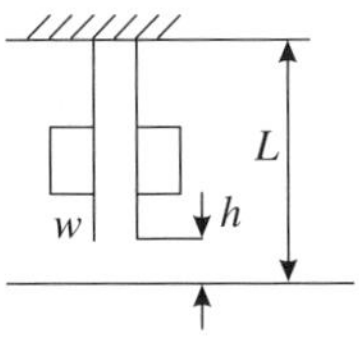

The maximum elongation of the bar is due to the impact of the falling weight w and the static elongation from this weight. With the assumption that the total work is converted into deformation we have the equation for total elongation $e = e_{st}(1+\sqrt{1 + 2h/e_{st}})$, where $e_{st} = w/ER$; the dynamic force depends on the velocity at the beginning of impact, $\upsilon = \upsilon 2gh$. The equation for elongation can be written as $e = e_{st}(1+\sqrt{1 + \upsilon^2/ge_{st}})$. If $\upsilon = 0$ (sudden impact) then $e = 2e_{st}$. If e is small in comparison with h, then $k_{dyn} = \sqrt{2h/e_{st}}$. The equation for total elastic deformation after substituting $e_{st} = w/ER$ becomes $e = \sqrt{2hw/ER}$, $de/dR = -\sqrt{2hw/E}/\sqrt[3]{R^2} = C_s$. The optimal geometrical stiffness is $R_o = \sqrt[3]{2hw/E} * 1/C_s$. The optimal area of cross-section is $A_o = R_oL$. The stress in the bar of optimal dimensions is $\sigma = 2hw/A_o < \sigma_y$.

10.2 Bending impact

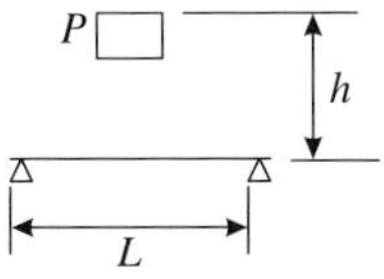

A simply supported beam is subjected to the sudden impact of load P that is falling from height h. The deflection of the beam in the case of impact is $Y_{dyn} = k_{dyn}Y_{st}$. The deflection from the dynamic force is equal to the static deflection from the force P times the dynamic coefficient $k_{dyn} = \upsilon 2h/Y_{dyn}$. In first approximation for sudden impact, $k_{dyn} = 2$. The optimal geometrical stiffness of the beam can be calculated with the equation of elastic stability, $R_o = \sqrt{M_{dyn}/EC_{st}}$, where $M_{dyn} = k_{dyn}M_{st}$.

Reliable design depends on the equation of elastic stability for all types of structures and forces. It is especially important for structures subjected to dynamic forces to establish the proper correlation of cross-section with length.

11. Testing of materials

The fundamental misinterpretation of the elastic relations in the linear theory of elasticity affects the testing of materials. Tests of materials are conducted on special specimens selected experimentally. The reason for this stated in the "Handbook of Engineering Fundamentals" edited by Ovid W. Eshbach and Mott Souders is as follows: "The fundamental data obtained in a test on material are affected by the method of testing and the size and shape of the specimen. To eliminate variation in results due to these causes, standards have been adopted by the ASTM, the ASME, and various associations of manufacturers."

The standards have been created for lack of scientific understanding. However the non-linear theory of elasticity provides scientific explanation of how size, shape and method of loading affect the limit for a specific structure. It provides the opportunity to test specimens of different sizes and shapes and to establish more accurate limits for materials. The geometrical stiffness of a specimen allows us to calculate the individual limiting stress for a specimen, $F_{cr}=ER^2C_{st}$, where the coefficient of elastic stability $C_{st}=4$. If the actual limit for this specimen is less than the calculated individual limit then it is a material limit. Illustration of this law is in the stress–strain diagram.

Appendix I. Optimal design of typical beams

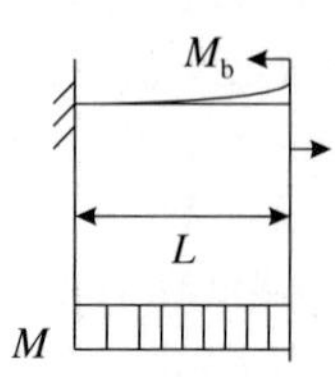

$M=M_b,\ R_o=\sqrt{M/E};\ C_s=1;$
$I_o=R_oL,\ \sigma_o=M/S_o\leqslant\sigma_{cr}\leqslant\sigma_y;$
$R_{cr}=0.5\sqrt{M/E};\ I_{cr}=R_{cr}L;\ \sigma_{cr}=M/S_{cr};$
$Y_{max}=ML/2ER;\ \theta_1=0;\ \theta_2=M/ER,$
$\theta=\theta_1+\theta_2.$

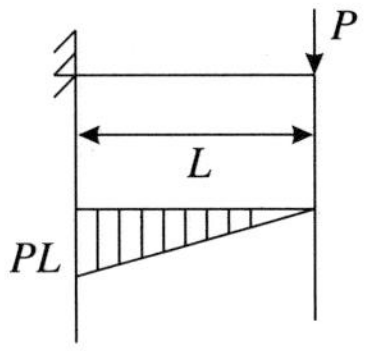

$M_{max} = PL$, $M = PL/2$, $R_o = \sqrt{M/E}$; $C_s = 1$;
$I_o = R_o L$;
$I_{cr} = R_{cr} L$; $R_{cr} = 0.5\sqrt{M/E}$, $C_s = 4$;
$\sigma_{cr} = M/S_{cr}$, $\sigma_o \leqslant \sigma_{cr} \leqslant \sigma_y$;
$Y = -PL^3/3EI$; $\theta_1 = 0$, $\theta_2 = -PL^2/2EI$.

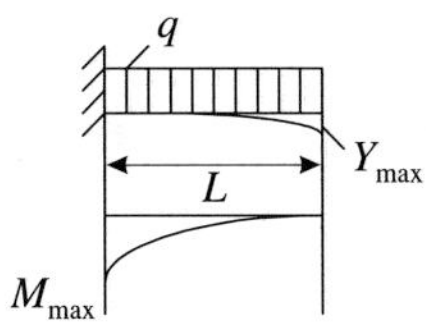

$M_{max} = 0.5qL^2$, $M = M_{max}/2$, $R_o = \sqrt{M/E}$;
$C_s = 1$;
$I_o = R_o L$;
$R_{cr} = 0.5\sqrt{M/E}$; $I_{cr} = R_{cr} L$;
$\sigma_o = M/S_o \leqslant \sigma_{cr} \leqslant \sigma_y$M;
$Y_{max} = -qL^4/8EI$; $\theta_1 = 0$; $\theta_2 = -qL/6EI$.

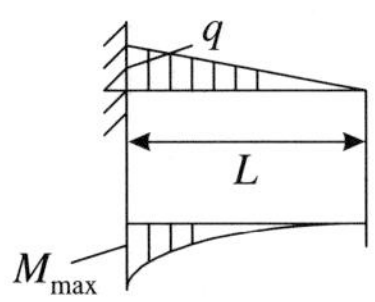

$M_{max} = \frac{1}{6}qL^2$, $M = M_{max}/2$, $R_o = \sqrt{M/E}$;
$I_o = R_o L$;
$R_{cr} = 0.5\sqrt{M/E}$, $I_{cr} = R_{cr} L$;
$\sigma_o \leqslant \sigma_{cr} \leqslant \sigma_y$, $\sigma = M/S$;
$Y_{max} = -qL^4/30EI$, $\theta_1 = 0$, $\theta = -qL^3/24EI$.

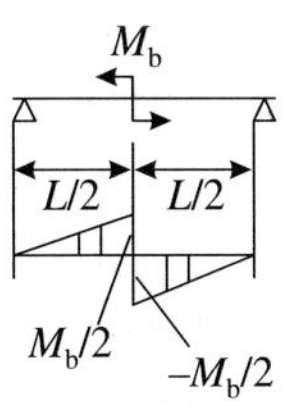

$M_{max} = M_b/2$, $M = M_{max}/2$, $R_o = \sqrt{M/E}$;
$I_o = R_o L$; $\sigma_o = M/S_o$;
$R_{cr} = 0.5\sqrt{M/E}$;
$\sigma_o \leqslant \sigma_{cr} \leqslant \sigma_y$;
$Y_{max} = -ML^2/128EI$, $\theta_1 = \theta_2 = -ML/24EI$.

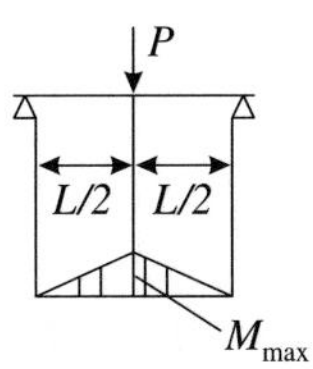

$M_{max} = PL/4$, $M = M_{max}/2$, $R = (M/EC_s)$;
$I_o = R_o L$;
$R_{cr} = 0.5(M/E)$, $I_{cr} = R_{cr} L$, $\sigma_{cr} = M/S_{cr}$;
$\sigma_o \leqslant \sigma_{cr} \leqslant \sigma_y$;
$Y_{max} = -PL/48EI$, $\theta_1 = \theta_2 = [PL/16EI]$;
$\theta = \theta_1 + \theta_2 = M/ER$.

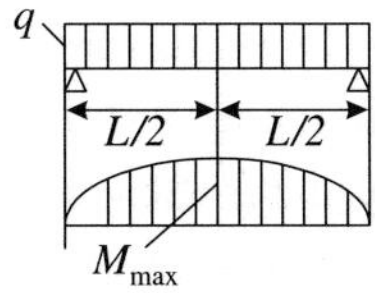

$M_{max} = \frac{1}{8}qL^2$, $M = M_{max}/2$, $R = \sqrt{M/EC_s}$;
$I = RL$, $\sigma = M/S$;
$\sigma_o \leqslant \sigma_{cr} \leqslant \sigma_y$;
$Y_{max} = -\frac{5}{384}(qL^4/EI)$, $\theta_1 = \theta_2 = [qL^3/24EI]$,
$\theta = \theta_1 + \theta_2 = M/ER$.

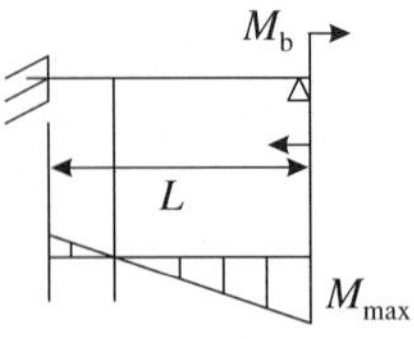

$M_{max} = M_b$, $M = M_{max}/2$, $R = \sqrt{M/EC_s}$;
$I = RL$, $\sigma = M/S$;
$C_{s,cr} = 4$, $C_{s,o} < 4$;
$\sigma_o \leqslant \sigma_{cr} \leqslant \sigma_y$;
$Y_{max} = -ML^2/27EI$, $\theta_1 = 0$, $\theta_2 = -ML/4EI$,
$\theta = \theta_1 + \theta_2 = M/ER$.

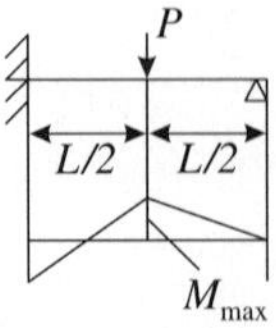

$M_{max} = \frac{3}{16}PL$, $M = M_{max}/2$, $R = \sqrt{M/EC_s}$;
$I = RL$;
$C_{s,cr} = 4$, $C_{s,o} < 4$;
$\sigma = M/S$, $\sigma_o \leqslant \sigma_{cr} \leqslant \sigma_y$;
$Y_{max} = -0.009PL^3/EI$, $\theta_1 = 0$,
$\theta_2 = PL^2/32EI$, $\theta = \theta_1 + \theta_2 = M/ER$.

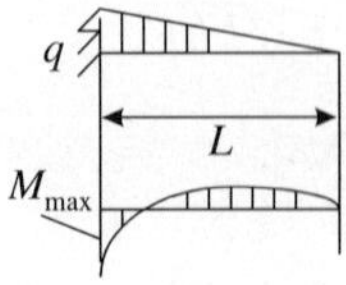

$M_{max} = \frac{1}{15}qL^2$, $M = M_{max}/2$, $R = \sqrt{M/EC_s}$;
$I = RL$;
$C_{s,cr} = 4$, $C_{s,o} < 4$;
$\sigma = M/S$, $\sigma_o \leqslant \sigma_{cr} \leqslant \sigma_y$;
$Y_{max} = -qL^4/418EI$, $\theta_1 = 0$, $\theta_2 = qL^3/120EI$.

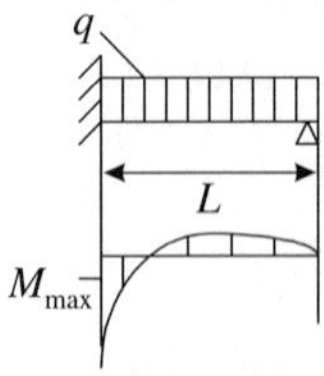

$M_{max} = \frac{1}{8}qL^2$, $M = M_{max}/2$, $R = \sqrt{M/EC_s}$;
$I = RL$;
$C_{s,cr} = 4$, $C_{s,o} < 4$;
$\sigma = M/S$, $\sigma_o \leqslant \sigma_{cr} \leqslant \sigma_y$;
$Y_{max} = -qL^4/48EI$, $\theta_1 = 0$, $\theta_2 = qL^3/120EI$,
$\theta = M/ER$.

Appendix II

Industrial products rarely come as a single design. Rather they are produced as a series of similar designs distinguished primarily by size. In such cases, each model in the series should behave in a comparable manner. Knowledge of the criterion of elastic stability allows us to reach comparable desirable behavior in a series of structures using a non-destructive test on the prototype or a model for the particular series. By measuring the deformation of a model upon applying an experimental force we can calculate the optimal dimensions for the model and for the similar structures. In the following we present the method of optimization that includes non-destructive test and simple programs for different types of structures.

Tension–compression

$\sigma = E\varepsilon = N/A$; $R_o = \sqrt{N/EC_s}$; $e=N/ER$; $R = N/Ee$; or $N = eER$; $R = kA/L$;

(1) Optimal geometrical stiffness, $R_o = \sqrt{A\varepsilon/C_s}$;
(2) Optimal area of cross-section, $A_o = AR_o/R$;
(3) Stress in the material of the model, $\sigma_a = A\varepsilon^2 EC_s/e$;
(4) Optimal geometrical stiffness for any structure in the set, $R'_o = R_o\sqrt{N'/N}$;
(5) Optimal area of cross-section $A'=R'L'$;
(6) Stress in the structure of optimal design should be checked against the limit of the material, $\sigma = N'/A'_o < \sigma_y$.

Bending

$e = M/ER$; $R = kS/L$;

(1) Optimal geometrical stiffness of the model, $R_o = \sqrt{\varepsilon S/C_s}$ where S is the section modulus of the model;
(2) Optimal geometrical stiffness for any structure in the set, $R'_o = R_o\sqrt{M'/M}$;
(3) Optimal section modulus, $S'_o = R'_o L'$;
(4) Stress in a beam of optimal design should be checked against the limit of material, $\sigma = M'/S'_o \leqslant \sigma_y$.

Circular cylindrical shells (membrane theory)

Axial tension/compression: r is the radius of the shell, t is its thickness.

(1) Optimal geometrical stiffness of the model, $R_o = \sqrt{\varepsilon A/C_s}$, where $A = 2\pi rt$ is the area of cross-section, r is the inside diameter, and t is the thickness of the shell;

(2) Optimal geometrical stiffness of a shell in the set, $R'_o = R_o\sqrt{N'/N}$;
(3) Optimal thickness of the shell, $t'_o = R'_o L'/2\pi r'$;
(4) Total tensile deformation of the shell, $D' = N'/ER'_o$;
(5) Stress in a shell of optimal dimensions should be checked against the limit of material, $\sigma = E\varepsilon < \sigma_y$.

Cylindrical shell under evenly distributed pressure Q and axial tensile force N:

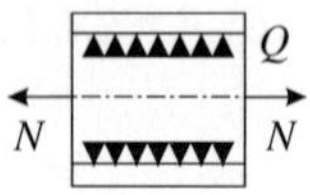

(1) The actual geometrical stiffness of the model is $R = A\varepsilon/D$, where strain ε and deformation D are measured along the cylinder;
(2) The equivalent tensile elastic force in the model is $N_{eq} = DER$;
(3) Optimal geometrical stiffness of the model, $R_o = \sqrt{A\varepsilon/C_s}$;
(4) Optimal geometrical stiffness for the shell, $R'_o = R_o\sqrt{N'/N}$;
(5) The equivalent tensile elastic force in the shell, $R_o^2/R_o'^2 = N_{eq}/N'_{eq}$;
(6) Optimal geometrical thickness of a shell in the set, $t'_o = R'_o L'/2r'$;
(7) Optimal geometrical stiffness in the radial direction, $R_{r,o} = \sqrt{tL\varepsilon_r/C_s}$;
(8) Compare the required values of geometrical stiffness in the longitudinal and radial directions and select the larger one, for instance, R'_o;
(9) Stress in the shell from the equivalent force should be less than the allowable material stress, $\sigma = N_{eq}/A < \sigma_y$.

Cylindrical shell closed at both ends, under internal pressure:
(1) Elongation/compression of the cylinder is $u = (1-2)QrL/2Et$; $R_u = 2t/L$; $R_{u,o} = \sqrt{(1-2\mu)Qr/EC_s}$; $t_{u,o} = R_{u,o}L/2$;
(2) Deformation in the tangential direction is $w = (2-\mu)Qr^2/2Et$; $R = 2t/r$; $R_{w,o} = \sqrt{(2-\mu)Qr/EC_s}$; $t_{w,o} = rR_{w,o}/2$;
(3) For optimal thickness t_o compare $t_{u,o}$ and $t_{w,o}$ and select the larger one;
(4) The stress in a shell of optimal thickness is $\sigma = (2-\mu)Qr/2t_o < \sigma_y$.

Round plate, simply supported, uniformly loaded:

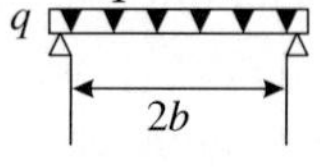

(1) Elastic moment is $M_{max} = (3+\mu)qb^2/16$.

(2) Geometrical stiffness is $R_o = \sqrt{M_{max}/EC_s}$.
(3) For the model $R_o = \sqrt{2bt\varepsilon/C_s}$; for any similar plate $R'_o = R_o\sqrt{M'/M}$.
(4) Optimal thickness is $t_o = \sqrt{6R_o}$; $R = S/2b$; $S_o = bt^2/3$; $R = t^2/6$.
(5) Maximum stress is $\sigma = M/S_o < \sigma_y$.
(6) Total deformation is $e = M/ER$.

Appendix III. Table for shaft calculation

In the following table, the geometrical stiffness of a shaft is $R = KI_p/L$; the equation for calculating the optimal geometrical stiffness is $R_o = \sqrt{M/EC_s}$; the maximum torsional stress is $\tau_{max} = M/S_p$; the twisting moment is $M_T = M$; the angular deformation is $\theta = M/ER$; the polar moment of inertia is I_p; the polar section modulus is S_p; and G is the shearing modulus of elasticity.

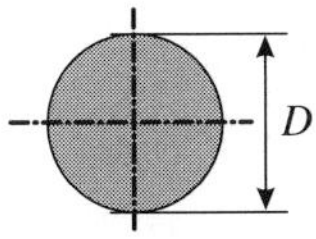

$I_p = \pi D^4/32 \approx 0.1D^4$
$S_p = \pi D^3/16 \approx 0.2D^3$

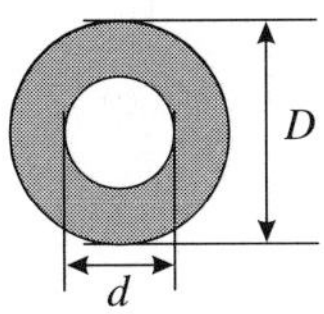

$I_p \approx 0.1D^4(1 - c^4)$
$S_p \approx 0.2D^3(1 - c^4)$
$c = d/D$

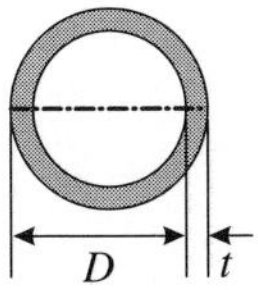

$I_p = \pi D^3 t/4$
$S_p = \pi D^2 t/2$

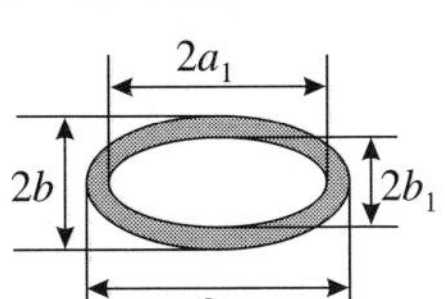

$I_p = \pi(1 - c^4)\,a^3b^3/(a^2 + b^2)$
$S_p = \pi(1 - c^4)\,ab^2/2$
$c = a_1/a = b_1/b$

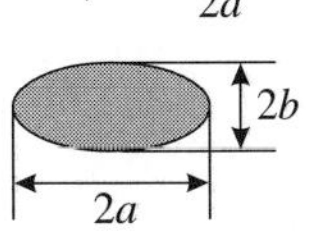

$I_p = \pi a^3b^3/(a^2 + b^2)$
$S_p = \pi ab^2/2$

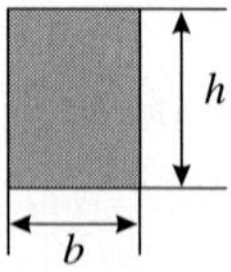

$$I_p = \beta h b^3$$
$$S_p = \alpha h b^2$$
for $h/b = 1$, $\alpha = 0.21$, $\beta = 0.14$;
for $h/b = 10$, $\alpha = 0.31$, $\beta = 0.31$.

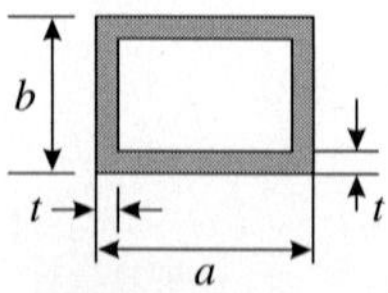

$$I_p = 2t_1 t_2 a^2 b^2 / (a t_1 + b t_2)$$
$$S_p = 2abt_2$$

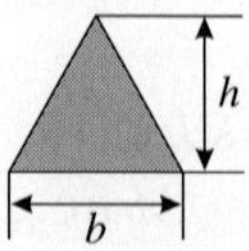

$$I_p = b^4/46 = h^4/26$$
$$S_p = 2I_p/h$$

Part IV. Further Discussions in the Theory of Elasticity

This Part presents on the one hand an extended consideration of the material that has been discussed already, and on the other hand a discussion of some other important problems in the theory of elasticity. These problems were not considered in Parts I and II, not because they are of secondary importance but rather for the purpose of concentrating attention on the central ideas necessary for the creation of our new theory and method.

1. Graph analysis

The common definition of a graph is "The graph of a function f is the graph of the equation $y=f(x)$." (Calculus, Howard E. Campbell and Paul F. Dierker). By definition the graph has taken secondary reflective place in mathematical analysis. If a function is correct then its graph is also correct. Here, in the process of building the non-linear theory of elasticity, the graph plays a role that is independent of and equal to that of the analytical description. The graph is separated from the analytical description of a function until these two independent approaches to the description of a phenomenon are merged.

We traditionally use graphs for:

(1) Visualization of the relationship between a function and a selected independent variable while other variables in the function are considered to be constants. For example, the graphical presentation of deformation versus geometrical stiffness, D vs. R, is a presentation of the equation $D=F/ER$ where the geometrical stiffness R is the selected independent variable. The elastic force F and the modulus of elasticity of the material E are the constants in this equation.

(2) Comparing the graphs described in item (1) for different values of the constants. There can be a number of curves in the same system of coordinates for different values of the variable-constants. This gives us the opportunity to examine the effect of a particular variable on the function. For example, the graph of deformation versus geometrical stiffness can be drawn for different tensile forces, $D=f(F_i/ER)$ for evaluating the effect of force on deformation and on the necessary optimal geometrical stiffness.
(3) Presenting the results of an experiment. For example, the graph can be drawn with an experimental device, such as a strain gauge. In the standard test on a material the device draws a curve of elongation for the different increasing forces applied to a specimen, $D=f(F)$.
(4) Identifying regularities or a law. The researcher can present graphically the results obtained on a number of different specimens in order to identify regular patterns or a law. This technique actually assigns the graph a more independent role. However, until mathematical proof for this graphical presentation is obtained the graphical presentation is as hypothetical as a mathematical hypothesis. Experimental data by themselves do not create a theory. The interpretation of data in connection with the logical–mathematical structure may result in a theory. For validating a theory one needs both analytical and empirical methods.
(5) Presenting statistical data obtained with numerous experiments or observations. Such a graph may show deviation of the actual data from the theoretical description or law. This statistical method may support a theory or disprove it.

In all these cases the graphical description is considered secondary to the mathematical description. There is no independent role ascribed to the graphical analysis in the process of developing a physical theory. And for practical purposes computational methods are more convenient and effective for getting the results. However, whether it be computation with formulas or reading from tables and diagrams, these methods belong to the established theory.

In the development of the non-linear theory of elasticity we used the graphical analysis independently and primarily for the building of the system of equations. And, actually, graphical analysis instrumental in conceiving the physical ideas. The graph was, also, used for verifying and correcting a function. The complete correspondence of the system of functions, consisting of a basic function and an independently known derivative function, to the graphical-geometrical presentation of the basic function gives confidence in the correctness of the description. Thus, the equation of deformation for tension in the linear theory of elasticity, $e = NL/EA$, and the derivative from this equation, $de/dA = -NL/EA^2$, have been tested prior to refuting the linear theory. The area of cross-section A has been selected as the independent variable. The graph e vs. A is constructed in order to understand the effect of cross-section on the deformation and on the elastic behavior of a structure. In drawing this graph we encounter a problem. If in the graphical presentation we select the scale for the absolute values of deformation the same as for the absolute values of area of cross-section then we will arrive at a distorted physical picture of elastic relations. The change of deformations that is, actually, the essence of the phenomenon of elasticity, will be missed completely in such a presentation because of the negligible values of deformation in comparison with the area of cross-section. The realization that it is essential to appreciate the changes of deformation at their own level led to the selection of equal scales for the significant figures of deformation versus the significant figures of the area of cross-section. Then, a curve e vs. A based on the experimental data will show changes of deformation due to the change of cross-section and at some point will show the elastic limit.

In order to reach a conclusion about the adequacy of the description we consider the correspondence of the description to the graph. To this end we compare the value of the tangent at any particular point of the curve with the derivative of the function at this point. For the traditional linear equation there is no correspondence between the graphical experimental presentation of the phenomenon and the analytical solution. Let us consider for example a simple beam of $L = 120$ in and $E = 30 \times 10^6$ psi, with a load $P = 6{,}000$ lbs at the center; then $M = PL/8 = 9 \times 10^4$ in-lb. The traditional formula for calculating the angular deformation is $\theta = ML/EI$. The graphical presentation of θ vs. I to equal scales with respect to their significant numbers suggests that at the point $\tan 45° = 1$ the significant figure for I is equal to the significant figure for θ. For example, the traditional equation for $I = 6\,\text{in}^4$ gives $\theta = (9 \times 10^4 * 120)/(30 \times 10^6 * 6) = 0.06$. While the graphical description of the relation between the significant numbers gives tan 45°, the equation of deformation at this point gives only $\mathrm{d}\theta/\mathrm{d}I = 0.01 = \tan 0.6°$. The great discrepancy between the results of the graphical presentation of the actual experimental relation and the results obtained with the equation does not suggest to neglect the experimental finding. It rather suggests to change the description. The new description is that the total deformation is proportional to the bending moment and inverse to the resistance of the beam to the bending, $\theta = M/ER$. In this case $\mathrm{d}\theta/\mathrm{d}R = 1$ in the equation will have an adequate graphical presentation for $R = \sqrt{M/E} = 0.055$, and $I = RL = 6.6\,\text{in}^4$. This gives a more correct graphical presentation of the experimental data that, also, corresponds to the analytical description. After verification of the system of equations the main method for solving practical problems becomes computational.

The foregoing example shows that the proper selection of the scales for the function and its independent variable is important in the process of evaluating physical relations and the results of an experiment. The next step is the proper selection of the independent variable and the correction of the function if necessary. We already discussed the selection of the independent variable. It was done under the condition that the function of deformation exist in the domain of the tangent function. In the interval of rapid changes the derivative of the function is in the interval of significant figures, $1 < \mathrm{d}D/\mathrm{d}R < 10$, so the deformation and the independent variable R can be presented graphically in the same scale. It helps to select a physical property that may have logically required values.

The analytical description should correspond to the graph presenting the real relations independently. It has been noted in the literature that "Often a formula may be rewritten in order to avoid a loss of significant figures." (*Mathematics*, by C. Edward Sandifer, from Marks' Standard Handbook for Mechanical Engineers.) From the same source: "Where it is possible, calculations and measurements should be planned so that loss of significant figures can be avoided." The rearrangement of the equation of deformation brought up the concept of geometrical stiffness as an independent variable.

The description of the elastic relations is based on understanding the phenomenon of elasticity. The basic concepts should relate to the phenomenon. The main observable entities of the phenomenon of elasticity in a structure are the elastic force, the resistance of the body to the change, and the change itself that is the deformation. The characteristic of resistance to deformation was described as ER. The variable part of the resistance depends mainly on the geometry of a structure. The value of geometrical stiffness has a dimensional level corresponding to the values of deformation. The geometrical stiffness can be presented in the same scale as the deformation. This correction of the equation of deformation gives reliable results both for the deformation and for the rate of change of deformation.

The rate of change of deformation describes the elastic behavior of a structure, and more specific, describes its elastic stability. The limit of elasticity corresponds to the limiting geometrical stiffness. The individual limiting stress depending on the geometrical stiffness was accurately predicted. The conclusion was that in order to obtain an adequate description of the phenomenon of elasticity one should select an independent variable at the same dimensional level as the function.

We can devise the equation of angular deformation and test it with an independently constructed graph of θ vs. R, where the geometrical stiffness is $R=KI/L$. Or, the equation of deformation can be written as $e=M/ER$, where e is the elongation of the external fibers of the beam. The geometrical stiffness of a beam in this description is $R=KS/L$, where S is the section modulus. Here in practice the construction of a reliable system of equations consists of adjusting a basic function to a known derivative function, which adequately describes the physical behavior or, in other words, describes the phenomenon.

The analysis of a graph presenting the function versus an independent variable plays an important and independent role in the selection of the variables and in the construction and verification of the analytical description of the phenomenon.

1.1 Commentary to Illustration 1 of Part I

Illustration 1 of Part I presents two different curves, both describing the phenomenon of elasticity. Although for both curves the physical variables, such as area of cross-section A, length L, force N and modulus of elasticity E are the same, and the function e has the same corresponding values, the curves show different behavior. In fact, Illustration 1 shows the curves in Illustrations 2 and 3 side by side. The purpose of this diagram is to show that for selecting an adequate description among the different possible descriptions of the same phenomenon it is necessary to examine the derivatives of the equations. One of the purposes of examining the derivative is the need to prove the hypothetical basic equation. The correspondence of the basic equation to a known derivative is the means to provide proof of logical correctness for the system of equations. Another reliable way of presenting the physical function is by identifying it with a known geometrical function that has the same pattern of behavior as the physical function and that at the same time is known to be true mathematically.

2. Geometrical models of physical functions

In Part I we attained verification of the initial hypothetical basic equation by the confirmation that its derivative is a known tangent function in both its intervals, i.e. in the interval of slow changes and in the interval of rapid changes. Also we discussed the reason behind this conclusion. There is no proof that the function is correct until the basic equation has a derivative that is known to be correct, such as the tangent function in the theory of elasticity.

We can as well select the basic equation as a known function, and then no adjustment would be necessary. Such a physical function would be proven by its design. For the description of deformation in tension, $e=N/ER$, an equivalent geometrical description can be made with the positive part of a hyperbola of which the asymptotes are parallel to the axes (see the graph below). The equation for such a hyperbola is $xy=a^2/2$. We may assign $y=e$, $x=R$, $a^2/2=N/E$. Then, the characteristic of the hyperbola $a=\sqrt{2N/E}$ transforms our equation of deformation into the known hyperbolic function, whose properties we can use for the selection of geometrical stiffness. The individual limit can be obtained with the equation of elastic stability.

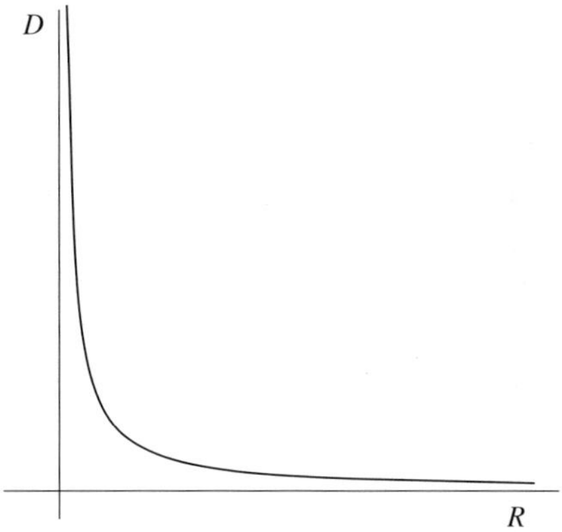

Thus, a reliable physical function can be devised by matching the physical function with a known geometrical function. Both functions have the same pattern of behavior. The hyperbolic characteristic a in the case of bending is $a=\sqrt{2M/E}$. The equation in its traditional form, such as $e=NL/EA$, cannot be identified with a hyperbola. Despite its appearance this function is linear within its physical limit. Such a description does not reflect elastic behavior correctly.

The important part of devising a theory is the selection of the variables, particularly the selection of the independent variable. The conditions for the selection of the independent variable are as following: the significant figure of the function should be comparable with the significant figure of the independent variable; the exponent 10^b of the independent variable should correspond to the exponent of the function.

The function or its derivative needs to be matched with a geometrical function whose logical–mathematical properties are known, apart from the possible physical content, as *a priori* conventional knowledge we agree with. The correctness of the description of the physical relations has then been provided and ensured mathematically. The basic equation should be, in addition, tested experimentally. It will confirm the factual correctness not only of the basic equation of deformation but also of the derivative equation for the rate of change of deformation that we cannot confirm empirically otherwise.

The graph shows a hyperbola whose properties can be identified with the equation of deformation.

3. The equation for the elastic line and the non-linear theory of elasticity

The function of deformation serves two different purposes. In the linear theory we deal with the description of deformations, displacements and stresses for an individual structure. Traditionally the calculation of the deflections along the elastic line is done with the Euler Bernoulli equation for the elastic line, $EI\,d^2y/dz^2 = \pm M$. The independent variable in this case is z, the distance from the origin of the coordinate axis to the particular point/section of a beam. The assumption is also made that the distribution of the internal elastic forces corresponds to the distribution of the external forces. Thus the equation for angular deformations is obtained as

$$\theta = \frac{1}{EI}\left[C + M(z-a) + \tfrac{1}{2}P(z-b)^2 + \tfrac{1}{6}q(z-c)^3 - \tfrac{1}{6}q(z-d)^3\right],$$

where $C = EI\theta_0$, and a, b, c and d are the corresponding distances to the applied forces. The equation for the elastic line is

$$Y = \frac{1}{EI}\left[D + Cz + \tfrac{1}{2}M(z-a)^2 + \tfrac{1}{6}P(z-b)^3 + \tfrac{1}{24}q(z-c)^4 - \tfrac{1}{24}q(z-d)^4\right],$$

where $D = EIY_0$. The formulas for determining the deformations and deflections in the particular cases, it seems, work satisfactorily for the purpose they serve.

The equation for the elastic line represents a physical body in schematic form. The derivative of this equation is the tangent to the elastic line that corresponds to the angular displacement of the cross-sections. Thus, it also describes physical changes in the body. The equation for the elastic line, $y = f(z)$, and the equation for the angular deformation, $\theta = f'(z)$, can both be tested experimentally.

In the non-linear theory of elasticity the function of deformation has a different assignment. The equation of deformation in the non-linear theory describes a set of similar structures. Each structure in the set behaves according to its place in the set. The other purpose of this description is to provide a mathematical–logical structure that is suitable for making physical prognoses and conclusions concerning the elastic behavior of a structure.

Thus the equation of deformation in the non-linear theory of elasticity is a mathematical curve describing relations rather than a physical body. The derivative of this equation predicts possible changes in the behavior of a structure rather than physical changes in the geometry of the body. This system consisting of a basic function and a derivative equation is a logical–mathematical system rather than the description of a body. It can be proven correct only by combining empirical/experimental verification with non-empirical/mathematical means of verification. The absolute values of the physical properties can be tested experimentally, but the values that describe relative changes, such as $d\theta/dR$, can be validated only in the context of a proven relation between the deformation θ and the geometrical stiffness R. The proof comes by validation of the two-level system of the functions. If the derivative function is a known function then the mathematically connected basic function of deformation is also logically necessarily correct. The empirical validation of the basic function proves both factual reliability of the equation of deformation and factual truth for the rate of change of deformation. Empirical and analytical truth go hand in hand.

There are no two truths and no two laws as philosophers have suggested.

Part V.
Philosophy and Logic of Physical Theory

1. Philosophical background of the non-linear theory of elasticity

> But if science may be said to be blind without philosophy, it is true that philosophy is virtually empty without science. *(A.J. Ayer, Language, Truth and Logic)*

> You ask me whether we need experience to know whether the definition of some attribute is true. To this I reply that we only need experience in the case of whatever cannot be deduced from the definition of a thing.
> *(B. Spinoza, On the Nature of Definition and Axiom)*

A new theory requires some explanation of the general principles on which it is founded. Here it is especially important because the philosophical principles of the non-linear theory of elasticity are not in the main stream of contemporary science. Nevertheless, most of these principles are not used rather than not known. It would be helpful to compare them with the current approaches and principles in science. The explanation is made in a manner connecting this particular theory with the general theory of knowledge. NLTE has its place in the philosophy of knowledge.

One of the general questions of epistemology concerns the sources of our knowledge in general and the sources of knowledge in physical theory in particular. In physics, adequate knowledge is intimately connected with the possibility to test and, ultimately, to prove a theory. Let us consider the treatments of this question in the two basic directions of epistemology, empiricism and rationalism. These directions differ significantly in their hypotheses on the sources of our ideas about the physical universe and on how ideas about the external world enter our mind. However, there is much less divergence between empiricism and rationalism on the methods of validation of the theories. Empirical and, in particular, experimental confirmation is viewed as the only ultimate source of proof or rejection of a physical theory.

The empiricist believes that the only reliable source of our ideas and, especially, physical ideas are observations of phenomena and experiments which stimulate understanding of the nature of things and help establish the relations between things. Our theories are the reflections of these understandings.

The great proponent of empiricism David Hume in his brilliant *An Enquiry Concerning Human Understanding* (1748) wrote: "All belief of matter of fact or real existence is derived merely from some object present to the memory or senses and a customary conjunction between that and some other object". And, "In a word, if we proceed not upon some fact present to the memory or senses, our reasoning would be merely hypothetical; the whole chain of inferences would have nothing to support it, nor could we ever, by its means, arrive at the knowledge of any real existence". And, "I shall venture to affirm as a general proposition, which admits of no exception, that the knowledge of this relation is not, in any instance, attained by reasoning a priori, but arises entirely from experience, when we find that any particular objects are constantly conjoined with each other" (D. Hume, An Enquiry Concerning Human Understanding).

The other trend in epistemology is the position of rationalism. According to rationalism we conceive ideas about reality intuitively, often from mental experiments, from primary ideas in our mind, and then deduce mathematical descriptions and test these experimentally.

"What is an idea? Thus the idea of things which exists in us is exclusively due to the fact that god, the author of both things and the mind, has endowed our mind with this power to infer from its own operation the truths which correspond perfectly to those of external things" (G.W. Leibniz, Knowledge and Metaphysics). A. Einstein seems generally to agree with this position: "The supreme task of the physicist is to arrive at those universal elementary laws from which the cosmos can be built up by pure deduction. There is no logical path to these laws; only intuition, resting on sympathetic understanding of experience can reach them". And again: "there is no logical bridge between phenomena and their theoretical principles; that is what Leibniz described so happily as "pre-established harmony" (A. Einstein, Ideas and Opinions, Principles of Research).

Rationalism cannot explain the connection between intuitive ideas and reality and ends up inevitably with experimental methods of verification as the only means to connect a theory to reality.

A material object represented by its properties and a phenomenon (something that happens to the object) is the substance of a physical theory. It is impossible to deny the importance of empirical methods in the creation and validation of a physical theory. However, our knowledge is not simply reflective. Theories are not discoveries of laws of nature, but rather laws created for a chosen phenomenon. Empirical methods by themselves are insufficient to create, support and much less to prove a theory.

Ideas can be conceived in different ways: by contemplating a problem with mathematics, by noting similarity of behavior of different things, or in by discovering patterns in the course of observation and experimentation. But no matter how the ideas come to us, the relations between things that we describe in a theory cannot be tested with empirical methods. When we know more about the structure of scientific theory and about the methods of testing the logical structure we have a better chance of creating a proven theory.

A physical theory establishes the relations between selected characteristics of a physical phenomenon. These characteristics can be 'sensed' or 'touched' and their quantities can be measured with their own physical units. For example, properties such as length, weight, energy, force, temperature, electric current, resistance, are measured in units of length, weight, etc. We connect the properties of a phenomenon by relations in order to predict these properties. The possibility to test and verify properties is considered to be the empirical truth. Geometrical forms and their descriptions are also part of the objective world. Geometrical parameters can be measured. However, geometrical theorems have properties that belong to the abstract conceptual category of relationships. Such a description is a logical structure that is self-consistent and intrinsically true. Logical structure does not need empirical verification. It can be considered for all practical purposes to be *a priori* knowledge as well as mathematical inferences. "Priori propositions are true if they are free of selfcontradiction" (Language, Truth and Logic, A.J. Ayer). *A priori* propositions describe the necessary unconditional relations.

Relations between things or properties are the objective part of the world. But the means of testing the relations are different from the means of testing the properties. A relation cannot be tested objectively with its own unit. There are no units for measuring relations. As an instrument for the objective identification of relations we can use the known patterns of these relations. Geometrical functions that represent logical-mathematical patterns known to be correct can prove or disprove the initial hypothetical relations. If a relation between physical characteristics can be described with a known function that is independently proven, then this becomes an objective proof of the particular relation and of all inferences from the description of such a relation. Convenient patterns for describing physical relations are trigonometric functions and geometrical descriptions such as circles, ellipses, hyperbolas, and parabolas. Geometrical functions, which have been proven, are part of our objective a priori knowledge of the physical world. This knowledge is non-empirical and the means of testing it is non-empirical as well. A reliable physical theory necessarily utilizes both empirical and non-empirical sources of knowledge and the corresponding means of testing.

Our ideas and reasoning concerning the relations between natural phenomena are much more complex than simply drawing conclusions from observations. Both methods, empirical and non-empirical, are equally objective sources of knowledge. Both methods necessarily should be used in the foundation of every adequate physical theory.

Here we do not speak about the subjective process of conceiving ideas by an individual. We speak about the objective necessity of the empirical and non-empirical parts of a physical theory: on the one hand physical characteristics and their empirical tests, on the other hand the relations between these physical entities and their non-empirical proof. A physical theory is the intrinsic unity of these objective complementary parts.

'Objective' also means that the process of constructing a theory can be formalized. The empirical part of a theory consists of the physical characteristics, which are the variables in the physical equation. These variables can be measured in units of the characteristics themselves. The relations between the physical characteristics must also be treated as real and objective. However these relations cannot be measured in their own units because this does not prove the objectiveness of the relation.

Relations are hypothetical until they are tested with something else representing known relations. Describing and testing physical relations can be carried out with the help of the patterns of relations that are mathematically proven functions. If our physical function corresponds to one of these known functions representing relations, then the continuous physical function can be proven correct for all of its points.

We build a theory using empirical and non-empirical methods of verification. Thus, the main difference between empirical and non-empirical properties of a phenomenon lies in the manner of testing the property. Both methods have a priori objective knowledge that allows us to draw objective conclusions.

The physical concepts of the non-linear theory of elasticity were selected after the mathematical pattern corresponding to the physical behavior was established. The physical characteristics are tested empirically, i.e., by measurement. The rate of change of deformation, on the other hand, is a mathematical property and it is proven mathematically by its identification with a tangent function.

Methods that are resting on intuition, faith in the perfect match of our ideas to reality and belief in our ability to deduce theory from observations have another thing in common, i.e., the necessity to make the mental leap from reality to theory. There is no logical bridge that can be constructed based on the principles of classical logic. Immanuel Kant, on the other hand, proposed a different path to knowledge and a different logic. "We must not seek the universal laws of nature in nature by means of experience, but conversely must seek nature, as to its universal conformity to law in the conditions of the possibility of experience, which lie in our sensibility and in our understanding" (I. Kant, Prolegomena to Any Future Metaphysics). The limitation that Kant sees is that "The object always remains unknown in itself; but when by the concept of the understanding the connection of the representations of the object, which are given to our sensibility, is determined as universally valid, the object is determined by this relation, and it is judgment that is objective". Currently we do not think "the object always remains unknown in itself".

But knowledge of the object itself does not necessarily lead to knowledge of its behavior and to knowledge of inter-relations. This is a different stage of knowledge. The description of relations and behavior does not belong to the empirical world, which is tested empirically, but rather to logic, which is tested with mathematical patterns.

Making a clear separation between the empirical content of a phenomenon and the non-empirical, but objective logical structure of the relations between the physical concepts allows us to build a logical bridge between the relations and the physical content of a theory. The selection of physical concepts has two sides to it. On the one hand, we select concepts to represent the real physical properties that can be tested experimentally, measured and verified. On the other hand, these concepts are the 'organic' parts of the logical-mathematical structure. We select the concepts to fit the structure rather than to satisfy our intuition or our preconceived ideas.

An empirical basis is important for conceiving some concepts. Thus, observation is helpful in establishing the connection between the cause that is an external force and the effect that is the deformation. However, this connection does not lead to a description of the functional relation between cause and effect. There is no such thing as a functional relation between cause and effect. The cause belongs to the external world while the functional relation belongs to the structure itself: it is the relation between deformation, elastic force, geometric characteristics and elastic properties of a material. Although the mean value of the elastic force can be equal to the mean value of the external forces, they are different forces. External forces with respect to a body are independent entities. The internal elastic force is the reaction of a body to changes in its geometry and therefore connects both to the geometry and to the changes in geometry.

In the process of constructing the non-linear theory the experiments were limited. One reason was that the author had no facilities and no means rather than that she doubted the usefulness of experiments. But there was also a clear realization of the fact that no amount of experiments will prove a physical theory and that only a limited number of experiments is needed to disprove a theory. Nevertheless, some experiments were performed. The first experiment on a number of specimens showed that the traditional formulas do not predict significant changes in the parameter range where drastic physical changes occur in reality. Apparently, it is impossible to find a limit with these formulas. Also, the conclusion was that there is no experiment that would allow the experimenter to test the rate of change of deformation over the full range of elastic relations. In contrast, the rate of change can be calculated with the new formulas and is often confirmed in the interval of rapid changes that demonstrates non-uniform changes. In a second experiment the individual limits for a number of specimens were calculated with the new equation of elastic stability and then compared with the actual limits. The correspondence between calculation and the occurrence of elastic failure is considered as support for the non-linear theory of elasticity.

The proof of the non-linear theory of elasticity comes not from these experiments but rather from the analytical verification of the relations. Empirical verification by itself is an insufficient source of proof or validation of a theory. There exist an uncounted number of experiments as well as statistical data collected over several centuries of engineering practice, which are not incorporated in a reliable theory. A trustworthy theory comes with a new logical structure of a theory and verification of this structure.

Physical theory is dual in nature: its physical part has its basis in empirical knowledge, while its logical-mathematical part has its own basis and its own means to prove the objective value of the relations established between the physical concepts. Part of the real world belongs to the relations and there is no objective empirically obtained data on the evaluation of the relations. We can measure the properties of the physical entities but not the relations between the properties. The empirical method by itself cannot provide us with complete objective knowledge of a phenomenon no matter how sophisticated our observational methods and how convincing the 'customary conjunction' looks. The importance of the empirical foundation for the physical sciences is realized, and there is no objection against it if it is considered as a part of a theoretical foundation. The other part is non-empirical. It is the part that provides a proof for a theory and may serve as a bridge between a phenomenon and its theoretical principles.

It is not necessary to leave the proof of hypothetical relations to our intuition. The relations can be evaluated objectively without involving empirical means: they can be tested with geometrical patterns representing certain relations. Thus, geometrical functions such as tangent, sine, cosine, etc., belong to the objectively known patterns of mathematical behavior. These functions can be used in physics for proving the objective truth of physical relations. The empirical and mathematical methods for verifying the relations intersect at some limiting points for the phenomenon. The realization that not everything can be tested experimentally and the separation of verification methods into empirical and mathematical methods makes the task of the researcher more clear and certain. The empirical method by itself cannot provide us with a complete knowledge of a phenomenon. The theory should meet the mathematical requirements of proof for its internal structure.

The other basic question of philosophy of knowledge is the cause–effect relation. "What is the nature of all our reasoning concerning matter of fact? The proper answer seems to be that they are founded on the relation of cause and effect" (D. Hume, An Inquiry Concerning Human Understanding).

The point of view here is that there is no functional relation between cause and effect. Therefore there is no reasoning based on this relation. Cause is always outside a system or a body. We can describe the functional relation only within the system, between the different properties of a body. It is within a whole or a body that one change is intimately connected with the other changes. The same cause, for example an external force, induces different responses of different bodies, for the response depends on the body. On the other hand, the same elastic force will evoke the same response from the different bodies for a certain condition, i.e., that the bodies have the same resistance to deformation. This is a functional relation that is considered as a law. The law describes only the effect, which is separated from the cause. Our reasoning concerns the functional relation and thus belongs to the effect rather than to cause–effect.

The relation between cause and effect is existential. It can be established as quantitative when employing an inductive method. The inductive method allows the forming of a reliable inference from some general law to particular relations. Thus, the general law of conservation of energy connects cause and effect for the phenomenon of elasticity. It lets us draw the conclusion that the work exerted by external forces is added to the elastic energy of the deformed body.

Yet another important question of philosophy concerns the relations between the whole and its parts. This question is important for physics, including the theory of elasticity. The whole can be a system, a body or part of a body depending on the purpose of the research. In all cases the whole is an entity that has parts. With respect to the properties of a whole we may say that the whole has more parts than the sum of its parts for we should consider the geometrical whole also as part of this division.

Thus, division of the entity with the purpose of finding some physical property includes consideration of the properties of the parts and the properties of the geometrical whole. In the non-linear theory of elasticity, in order to find the strength of a body or the limit of elasticity, the geometry of the body is separated from its material content. A property such as the limit of elasticity of a body is a relative characteristic. It can be the limit of the material or it can be the limit generated by the geometry of a structure. From a philosophical point of view the whole can be presented as consisting of a geometric form separated from its material content and the material content. Division of a whole into its constituting parts is one stage of abstraction for the purpose of acquiring knowledge of the properties of the parts. Separating the form from the material content is another stage that allows us to compare the characteristics of the material to the characteristics of the geometry.

2. Logic and physical theory

> There is no absolute standard of rationality, just as there is no method of constructing hypotheses which is guaranteed to be reliable
>
> *(A.J. Ayer, Language, Truth and Logic)*

2.1 Role of logic in science

There still is no consent about the subject matter of logic. Some scientists have the point of view that logic is the science of the necessary laws of thought processes. A more common contemporary view is that logic is the science of ordered objective relations independent of thought. Our point of view is that logic is the science dealing with the evaluation of objective ordered relationships as truth or falsehood.

Logic is a necessary part of constructing and proving a physical theory. The construction of a physical theory is a multifaceted endeavor. It has been proved that it is impossible to prove a theory only by physical means, i.e., with observations and empirical data. It has also been proved that it is impossible to prove a mathematical statement by means of mathematics. Kurt Gödel in his incompleteness theorem proved this impossibility. This also means that a physical theory that uses mathematical descriptions and mathematical deductive inferences cannot be proved mathematically. The general epistemological conclusion here is that a system needs an apparatus for its evaluation that is situated outside the system.

The science that can answer the question whether our theoretical constructions are true or false is logic. It can pass a judgment on both the physical and the mathematical part of a physical theory. The tools of logic are particular methods of reasoning that ultimately bring certainty to a theory. These methods lead from some hypothetical idea about a phenomenon to the establishment of a connection between cause and effect, to the selection of the characteristics describing the phenomenon, and to building and testing statements that describe the relations between those characteristics. Functions are used to predict the behavior of the object of investigation. These predictions are also needed in scientific research and in solving engineering problems.

The current common definition of logic is quite vague. For example, "Logic is the most general of all the sciences; it deals with the elements or operations common to all of them." (A Preface to Logic, Morris Raphael Cohen). This definition reflects uncertainty on the role of logic in science. The point of view that logic provides rules and elements of operations common to all sciences seems to have much in common with the definition that can be given of mathematics, and provides no specifics for distinguishing between logic and mathematics. However, these are two different scientific disciplines with different purposes and methods. Also, at this stage of development of logic in sciences, it would be more useful to have more specific approaches for different sciences. Then it would become possible to construct the logic of common elements and procedures. Here we consider the rules of logic in a physical theory, particularly in the theory of elasticity. Our conclusion is that logical analysis should be part of the construction, testing and proof of every scientific theory. Only such analysis can give the necessary degree of confidence in the theory.

Logic in physics is different from the current logic in mathematics. A comparison shows both similarities and significant differences. For example, any mathematical function is correct, and any mathematical inference from that function is correct. This approach is improper in physics. Only one of a number of possible descriptions is correct for each physical system. The physical content of a mathematical function places restrictions on the mathematical transformations. The rules of logic that can be applied to physical functions and procedures allow us to reach the necessary exclusiveness by justifying the physical function and mathematical procedures.

There are two main methods of logic that can be used for justification, i.e., inductive and deductive methods. These methods differ in their ways of making transitions from proposition to conclusion. In order to make such transitions successful we had to change the definition of inductive logic. The current common definition of an inductive process refers to the transition from accumulated empirical data to a conclusion. Here the definition of inductive logic is applied to the transition from general laws and definitions to a particular law that falls under the umbrella of the general law.

Mathematics considers inductive inference as a transition from "next to next" in the order of natural numbers. In physics a definitive inductive transition can be made only from general laws to specific or partial conclusions and laws. Mathematics provides constructions that can be useful in deductive logic. Physics has its own useful constructions such as inductive inference that can also be employed in general logic. I believe that chemistry, biology and medicine have their own very important logical rules that can be utilized in the general logic.

Looking back from the position of the constructed non-linear theory of elasticity one may see the necessity of including a logical system in a physical theory. A logical system consisting of a propositional function and a derivative function that leads from the selected initial data to a unique definitive mathematically provable prediction is a requirement for a physical theory. And although it is impossible to prove an individual statement, it is possible to prove the conjunction of the interdependent statements in the common domain of the physical phenomenon.

However, this logical structure and the method for proving its connection to the physical observations are not readily available in the literature on logic. Furthermore, classical logic and epistemology reject the possibility of logical structures that can be proven as true to the facts. However, it has been shown that NLTE satisfies the demand of certainty and provability. This theory has a logical structure that is consistent in all of its parts. We may therefore view the principles of this logic as an important addition to classical logic. We call it tentatively *definitive logic*.

Thus, in our view the role of logic in science is in the application of the rules of inductive and deductive inference of definitive logic for justification of all terms, mathematical structures and conclusions of a theory.

2.2 General argument

Let us consider a general argument in different logical systems. Classical logic has the following well-known argument at the foundation of its reasoning. "If these facts are admitted, then this conclusion must be admitted" and in more concise and formal way "If P, then Q". In modern deductive logic the definition of a valid deductive argument, which connects a propositional statement with the statement of a conclusion, is that "An argument is defined as a valid deductive argument if its premises could not all be true while its conclusion was false." (Modern Deductive Logic, Robert J. Ackerman).

The difference between the classical argument and this modern variation is that the argument "If P, then Q" has direction. It is asymmetrical from P inferring Q. The argument in modern deductive logic is an attempt to correct this incomplete connection of the classical argument, which does not lead to proof. At the same time, the modern argument by negating a conclusive deductive construction is telling us that we cannot prove a conclusion, we can only disprove it. The positive, but incomplete argument of classical logic versus the negative approach in modern deductive logic gives us no choice for building a successful logical structure.

In all those definitions of the validity of an argument it is assumed that premises and conclusions can be verified separately as true or false, and then we may transfer the conclusion to the connected statement. It is also assumed that the empirical judgment provides the sufficient and the only means of justification. The logical structure (proposition and conclusion) is tested by empirical validation of the statements. It is often assumed that if the terms of the propositional function are the correct experimentally obtained values and the function has an experimentally confirmed value, then not only the proposition but also the properly obtained mathematical inference must be true. This would be the case from the point of view of mathematics, but not in physics.

As has been discussed to some extent in Part I, empirical support is not a logical proof of a proposition, neither a proof of the conclusion. It should also be noted that logical falsification of the system, if the proposition is empirically correct, is no more easily attainable than logical proof of the whole system. The fact of the matter is that empirical means are not the only means that we use for validating or falsifying a physical theory. Not all terms and statements of a theory can be tested empirically. For example, it was shown earlier that the inference from a propositional function describing relations within a set of similar structures, in principle, cannot be measured and tested empirically, for we deal with logical–mathematical relations.

The process of building the non-linear theory of elasticity raised different questions than those we traditionally consider in logic. Thus, the selection of terms, construction of the propositional function, and connection of the propositional and derivative function, requires the application of logical methods. Furthermore, it appears that verification of physical relations cannot be achieved by empirical means only. We also need logical means for verification of physical relations. In traditional logic the opposite view is rather common, i.e., that logical constructions require experimental validation. However, in constructing NLTE we came to the conclusion that while some terms and statements can be verified empirically, without additional logical verification there is no proof for consistency of a physical theory and no proof of existence in reality of the relations that we describe. It is necessary to judge truth–falsehood of terms, statements and the system of statements, and to judge the theory in its entirety. Such judgment includes both empirical and logical means of validation.

The logical structure that is applicable to physical phenomena necessarily has two parts. It is the system of a propositional function and a derivative function. The propositional mathematical description of the relations between the terms needs experimental and logical validation of its terms. Hence in the equation of deformation, $D = F/ER$, each of the observable terms referring to the geometry is measurable. The theoretical term in this function, i.e. the elastic force, on the other hand, can be justified only by an inductive process. Inductive reasoning gives us secure inferences from universal laws and definitions to specific terms and statements of the theory in question. The non-observable theoretical term of elastic force is found by emploiting the definition of work. The elastic force is non-observable, nevertheless it is an objective part of the described relations in the concrete system of knowledge.

The value of deformation in the propositional function is empirically validated by measurement. This does not yet mean that the deformation function is necessarily correct. The same results can be obtained with different descriptions. These are equivalent from a mathematical viewpoint, but not in physics. Physical relations are unique and can have only one adequate description of the relations between the terms proven to be consistent. Only a logical deductive system of equations and its inductive validation in the system of scientific knowledge may prove a theory.

Having constructed the system of equations we need a method that allows us to test the system and eliminate the existentially incorrect propositions. We cannot declare the system of functions to be correct if we test the propositional function empirically and find correct values. In mathematics any function is considered valid and inference from it should be valid as well. But in physics we need to test more than the mathematical correctness or even the empirical correctness of the proposition: we also need to test the logical correctness of the relations. Inference has some pure mathematical–logical properties. The result of mathematical inference describes the rate of change of a function. We cannot test the purely mathematical properties experimentally. The system of a propositional function and its derivative can be proved or disproved as a system that needs both empirical and logical means of verification. The logical test of the system has to show that it is free of contradiction, is self-consistent, and is in the domain of possible physical existence.

The derivative equation describes the rate of change of a function. For a continuous parabolic function that has intervals of slow and rapid changes the derivative function has the character of a tangent function. The system of those functions can be correct only if both functions have the same domain. This inductive conclusion in NLTE validates both functions. Then the existential validity of the derivative function depends on empirical verification of the propositional function.

Suppose that Q is a statement obtained as a mathematical inference from statement P, and that these are connected with the necessary mathematical deductive process. Then the statement "If P then Q" abstracts a true assertion if and only if both P and Q are true assertions and have the same real domain. If some universal function or law covers inference Q then it would validate P and Q. Thus, the logical conjunction of P and Q is necessarily true only in a common domain with some universal function. In order for the system to be justified each of the statements need to be true. The difference is in the character of proof. A conclusion is a statement obtained deductively and validated inductively. Premises are constructed and proved with empirical and inductive validation. *A deductive–inductive system consisting of proposition and inference is logically justified if it is in the domain that validates the possibility of its physical existence. In a logically justified system, if P is empirically validated, then so is its derivative Q.*

3. The rules of logic

For building a physical theory we need not only physical ideas and hypotheses concerning the phenomenon, but we also need the rules of logic for the justification of every step of the process. The rules should correspond to the problem we have to solve. These problems are:

(1) The first problem concerns connecting the cause of a phenomenon with the effect. As a matter of fact, the cause finds its realization in the particular phenomenon that is its effect. The phenomenon becomes a logical structure designated as "effect". Observations may give us ideas that would help in separating cause from effect. But these observations cannot help us in establishing the logical connection between cause and effect. Classical logic and for that matter modern deductive logic do not offer a method for establishing a logical connection between cause and effect. Furthermore, epistemology doubts even the possibility of establishing such a connection.

(2) For constructing a reliable theory we also need logical rules for selecting the characteristics of a phenomenon that can be justified and used in the theory. There are no such rules in classical logic.

(3) There are no logical rules for constructing a propositional statement. In fact, no clear definition of a propositional statement seems to be available. The main argument of classical logic is formulated as a transition from data to conclusion. That definition is confusing; it may justify an incorrect approach to the building of a logical structure. For example, assume that the proposition in the linear theory of elasticity is the data representing force N, length of a bar L, area of cross-section A and modulus of elasticity of the material E. A mathematical statement is commonly believed to be a logical structure that allows the transition from data to conclusion. Then the conclusion regarding deformation of a structure can be reached with the well-known function $e = NL/EA$. Only an empirical test is needed to support this logical structure. Here, classical logic seems to fulfill its role in constructing the linear theory. However, analysis of this statement shows that it is an incomplete, inconclusive description of the elastic relations. It fails to indicate the position of a specific structure in a set of similar structures. One can see the logical insufficiency of such a construction in the inconclusive character of this statement. Many different statements can use the same initial data for reaching the same deformation result, but this does not mean that we deal with equivalent descriptions. There are no truly equivalent descriptions in physical theory. We need a logical method for selecting the

adequate description. Both classical and modern logic have no rule that requires logical selection and confirmation of a statement.

(4) The main argument of classical logic can be understood as a transition not from data to conclusion but from propositional statement to concluding statement. This would provide us with a complete logical structure. Such unique structure is a logical necessity.

(5) Now we have to test the logical structure. Classical logic has no definitive rules for testing it. Empirical validation is considered as the only means of testing a physical theory. But not all terms of a physical statement can be tested experimentally. Non-observable terms need logical justification. Classical logic does not provide a method for that. It also does not realize that the logic of relations cannot be justified empirically. We can test the basic propositional function experimentally. This test may validate that the quantitative relation in the function gives a satisfactory result for the deformation, but it cannot confirm the true–false nature of the relations among the components. The function still remains hypothetical.

(6) Making the inference is a necessary logical step. This step is usually missing in scientific theories. It is absent from the linear theory of elasticity as well. But even if we infer from the propositional function a mathematical conclusion, this new function inferred from a hypothetical function would also be a hypothetical function.

The derivative function, which is supposed to be an ultimate conclusion according to the methodology of classical logic, also needs to be tested against facts, for empirical facts are considered to be the only appropriate means for verifying a physical theory. This test is impossible for a logical structure.

(7) We need to make a distinction between purely mathematical proceedings and logical proceedings. In a superficial approach these proceedings seem analogous. In fact they are not. On the example of Euler's description of the elastic line we already discussed that the derivative of the equation of the elastic line is mathematical rather than logical inference. It does not tell us much about elastic relations, and it justifies anything. The derivative equation in this case describes the elastic line in terms of the angular values of the tangent to the line, which can be measured. On the other hand, the derivative of our log-

ical proposition, which shows the elastic relations depending on the position of a structure in a set of the similar structures, while obtained mathematically in the same fashion, has a purely logical purpose and property to it. One may see the difference as similar to the two-fold meaning of a geometrical function that has both physical meaning and logical meaning depending on its use. The main explanation for the necessity of a logical method for testing the logical derivative function is that its result is a relative mathematical value that cannot in principle be tested empirically. Empirical validation of the derivative function is possibly only in the interval of elastic failure.

In summary of this discussion we may say the following: The logical argument "If proposition P is correct, then conclusion Q is also correct", gives rise to questions and objections. What do we mean by "proposition"? Does it refer to the selected facts or to relations between these facts? How are the facts selected? What proof do we need for the assertion that the propositional equation is true? What proof do we have that we properly selected and connected the facts? Would it be sufficient if facts confirm the results of calculations of the hypothetical equation?

In traditional logic, verification is the process of confirming a hypothesis by showing that its predictions agree with the facts. We have to confirm the basic propositional function and the derivative function. Does every propositional function have a meaningful derivative? Can every function be confirmed empirically? This question arises because not every equation has measurable values as its result. The derivative equation in the non-linear theory of elasticity describes a mathematical relation between the members of a set of similar structures. We have no physical means to test the mathematical–logical relation. On the example of the theory of elasticity it was shown that inference from a basic function that is empirical in nature can be a logical function. The conclusion is presented as the ratio of two physical variables. We have no physical means to measure the relation and to objectively evaluate its existential value. If we try to do that then in our evaluation we again make an assumption: the result is correct if we selected the variables correctly. This is a circular logical procedure.

Empirical support of a function is necessary but not sufficient for asserting that a function is correct. There is no method in classical logic for proving the system of assumption and conclusion. However, validation is possible for we deal with a logical system rather than with the separate logical statements of proposition and conclusion. Classical logic has no answer to the question what designates the truth of a logical system and how to reach it.

The main achievement of deductive logic is the acceptance of the mathematically necessary conjunction of a propositional statement and its derivative as the basic logical structure of a physical theory.

A different logical system is implemented in the non-linear theory of elasticity. The possibility of attaining the correct final result on the inquiry about the elastic relations depends on the logical structure of a theory. *The logical structure of a theory should precondition a single result that is proven to be correct*. One of the rules in evaluating the logical structure is that the proposition describes the relations for a whole object and a set of similar objects under consideration. Another requirement is that the logical system should be complete, i.e., has a propositional statement and a deductive concluding statement. It is impossible to prove only the assumption or only the conclusion. From the point of view of definitive logic it is wrong to say "If the premises are correct, then the conclusion is correct". Premises are the part of logical structure that can be proved only in conjunction with its conclusion. *For proving a logical structure we need empirical confirmation of the premises, logical confirmation of the conclusion, and logical confirmation of the domain for the system of equations*. The premises can be verified empirically, but they lack logical validation. The conclusion, on the other hand, can be proven logically but by itself has no empirical validation.

The logical structure in physical theory is presented with a system of interconnected functions: a basic equation and its derivative. We start a theory with some physical idea/hypothesis, then construct the basic equation, which is empirical in nature, then obtain the derivative statement. The next necessary step is the consolidation of the derivative equation with the observed phenomenon. Both functions, proposition and conclusion, should be true to the observed phenomenon. This can be attained only in the domain of physical existence of the phenomenon.

Not all features of a phenomenon are measurable. The result of a derivative equation is a mathematical relative value describing the relation. We cannot assume that this relation is correct unless a universally known consistent mathematical function describes the relation adequately. What we can and should do is to find an analogous mathematical relation that has quantitative relations known to be true. If the derivative function is known and it is in the domain of possible physical existence, then the corresponding basic equation is necessarily logically connected to the correct structure. If we find a universal function that corresponds to the derivative function then we prove mathematically not only the derivative function but the corresponding basic function as well. The proofs of these functions have different nature. The basic function, which is empirical in nature, describes the quantitative relation between the physical concepts/characteristics of phenomenon; these are present in the equation as physical entities. This function should be proven empirically. The derivative function, on the other hand, can be proven as the inductive inference from a universal geometrical function rather than as a derivative from the hypothetical proposition.

This system of functions needs both empirical and logical verification if it is to be accepted as true. To that end the domain of the physical function should be established with an inductive procedure related to the universal function and to the physical domain of the phenomenon.

The universal function of which the properties indicate the domain of stable physical relations is the tangent function in the approximate interval $0 < \tan\alpha < 4$.

4. Logic of construction in NLTE

The system of definitive logic can also be laid out in the methodological steps of research. As an example of such procedure we consider the non-linear theory of elasticity.

(1) First, we have to outline the area of research, the object of research and the conditions. This process is connected with observations. Here we deal with a solid body that has supports and has contacts with other bodies or media representing external forces.

(2) A generalized description follows the observations. All solid bodies and all materials possess elastic properties within some limit.

A body changes its dimensions and shape under the action of an external force. The external force that is the cause of deformation of the body remains preserved in this interaction.

(3) Separation of cause from effect is the next step of logical analysis. The deformation and stress in a body are effects of interaction. The cause of deformation is the application of external forces to a solid body that has one or more supports.

(4) The investigation concerns not the cause, but the characteristics of the changed body, for this knowledge may help us to design strong and reliable structures. Our tendency to connect cause and effect is psychologically understandable. We see the cause (in most cases) and we see the subsequent effect: most obvious changes in the body. It is our empirical attitude to the world that urges us to include cause in a mathematical description. In fact, we are already account for the cause as it is absorbed by the body and represented as an effect. For understanding what happens to a body we have to concentrate on a body that is changed already as the result of application of an external force. The law of changes in a body belongs to the body itself. A body that we consider is a whole rather than the sum of its infinitesimal parts.

(5) The deformation D of a body depends on the resistance of the body to deformation due to its geometry R and the elastic properties of the material. Deformation is inversely proportional to the resistance. The elastic force F acting in the body is proportional to the deformation. The methods for obtaining the values of R and F are different. We can measure the geometrical stiffness, which is an observable compound term. We cannot measure elastic force, which is a non-observable theoretical term. The elastic force can be obtained using an inductive process that involves the general definition of work and the law of energy conservation. We assume that the work of an external force is stored in the body as elastic potential energy. The reverse job, that is the complete restoration of the initial geometry and position of a body after removing the external force, is then equal to the work performed by the external force. Thus the magnitude of the elastic force is equated to the magnitude of the external force. However, for determining stress in

a body we have to know not only the magnitude of the elastic force but also its distribution in the body.

(6) The description of the elastic relations is still the hypothetical statement $D = F/ER$. This function has observable and non-observable terms. The validation of the terms involves inductive and empirical methods. The assumption is made that elastic forces are proportional to the deformation. This assumption is described with Hooke's law of equivalence of stress and deformation, $\sigma = E\varepsilon$. The logical assumption from Hooke's law is that the distribution of elastic forces follows the distribution of deformations.

Another assumption was introduced in the linear theory of elasticity, i.e. that the elastic forces are in static equilibrium with the external forces and that the distribution of the elastic forces follows the distribution of the external forces. The latter assumption, which prevails in current practice, has less logical–mathematical validation for it contradicts Hooke's law, which is the main assumption on which linear and non-linear theories are built.

(7) The conclusion from the analysis of the equation of deformation is that we cannot establish whether the function is logically and existentially correct. Experimental validation of an equation does not prove that the function is correct. Different functions can be constructed using the same physical presuppositions but represented with different physical concepts. They can produce the same results that can be confirmed experimentally, but only one of them can be true. Empirical truth is insufficient for validating the relations in a proposition.

(8) Although each element of a function has physical meaning, such as elastic force, geometrical stiffness, modulus of elasticity and deformation, at the same time these concepts are different abstractions of reality. The invisible elastic force exists in the system of classical mechanics. We made that inductive logical inference from the fact that work is performed when external forces are removed. However we cannot measure this force directly. Its distribution is yet hypothetical. The magnitude of the elastic force can be found analytically. The coefficient of elasticity E describes a physical inherent property of a material. We made this assumption

because of the fact that in experiment different materials have different stress/strain ratios, $E = \sigma/\varepsilon$. We need a coefficient for equating the different properties of an elastic field. The coefficient is a hypothetical physical abstraction that can be found analytically only. The new concept of geometrical stiffness of a structure has its physical meaning though it is an abstraction that detaches the geometrical form of a body from its material substance. It can be measured in a specially designed experiment or it can be found from the equation of deformation and the equation that describes the relations between the geometrical parameters of a body such as $R = A/L$. The overall deformation of a body can be measured or calculated. The relations among these abstract and more or less hypothetical concepts are not proven and cannot be proven by empirical means. Also, as there are no reliable empirical means for measuring these relations, we need an independent logical proof of existential validity of the described relations.

(9) The most accurate description of the relations can be obtained mathematically with a partial derivative equation. As the basic equation is hypothetical, the derivative equation is also hypothetical. However we can find an analogous mathematical relation in the form of a theorem of plane geometry that is logically true and consistent by its design.

(10) The basic propositional function and its derivative should occupy the domain that is characteristic for stable physical relations. Stable elastic relations are characterized by the rate of change of the function in the interval of the tangent $0 < \tan\alpha < 4$.

(11) The rate of change of deformation in the case of dynamic forces changing in time should conform to the same law of elastic stability of a structure that is defined in the approximate interval, $0 < \mathrm{d}y/\mathrm{d}x < 4$, in case of a static force.

5. The definitive logic

We should recognize that conceiving the general physical idea describing the phenomenon is the most difficult and creative part of a theory of elasticity and, probably, of any theory. In the theory of elasticity it is the comparison of a body with a spring that started the theory. For millennia man had observed and used the springing qualities of solid bodies without making this connection to the 'spring'. Observation, though important, is not the most stimulating factor of a creative process. We cannot absolutely formalize the process of acquiring a physical idea. However, singling out a phenomenon, formulating the problems connected with the phenomenon, and separating cause from effect may help to recognize the basic physical concepts of a phenomenon. Concerning the formal part of a theory we can say that it is intimately connected with the definitive logical process. Thus, in the non-linear theory of elasticity this process can be presented in the following steps:

(1) Separating the cause, i.e. an external force, from the effect, i.e. the deformation of a body. The theory of elasticity considers not the external force, which is an independent entity that does not change in the process of deformation, but rather the deformation of the structure, for it is our purpose to design a reliable structure.
(2) Selecting the main characteristics related to the phenomenon as a whole and to its infinitesimal parts. Such characteristics are deformation, elastic force, and geometric characteristics of the body, and modulus of elasticity, stress, and strain.
(3) Establishing the existential status of the terms, i.e., which terms are observable and can be measured and which terms are theoretical and need logical justification.
(4) Evaluating the units of the terms/concepts in the description, including the consideration of the order of the units for correspondence of the order of the independent variable to the order of the function.
(5) Recognizing the three types of relations which require different logical approaches.

The equation of equivalence concerns the basic conceptual relation in the infinitesimal parts of a body, i.e. the relation between strain, an observable term, and stress, a theoretical term. The connection can be established inductively. The work of an external force is stored as elastic energy of the body for the whole body and for each of its infinitesimal parts. In simple cases the relation can be supported experimentally. In more complex cases empirical validation is necessary for testing the hypotheses of stress distribution.

The equation for the elastic line describes physical changes in the geometry of a body. The derivative dy/dz also describes physical changes along the elastic line. The connection between these two descriptions is mathematical rather than logical. These two equations, i.e. the equation of displacements and the equation of angular deformation, can be verified empirically.

The third type of relations pertains to a set of similar structures. We deal with more complex logical decisions in establishing the relations in a set. The logical structure necessarily consists of a basic equation and its derivative. There are certain logical demands for this structure: the derivative equation, or the basic function, should correspond to a self-consistent function such as a geometrical function with established properties. Also, this system has to be in the domain of stable physical existence. Thus, it is connected to physical necessity.

(6) The domain is resolved inductively. More specifically, it is based on the known character of the tangent function. The limit of stable physical relations corresponds approximately to the rate of change $[dy/dx] \leqslant 4$, or it is in the interval $0 < \tan \alpha < 4$. This is a necessary conclusion as the rate beyond this limit rapidly increases toward infinity, incompatible with the stability of physical relations.

(7) Thus, the complete logical system consisting of the propositional basic equation and its derivative equation should exist in the domain of stable physical relations. For the construction of the system of functions we may start with the simplest possible case: application of a static force and the corresponding static deformation. For selecting a structure with desirable elastic properties we consider a set of similar structures. The description of the elastic relations for the members of the set is the same for every individual structure in the set, i.e. the deformation is proportional to the elastic force and inverse to the resistance of the body to deformation. The elastic force in this case is constant. Then the independent variable is the resistance of the structure due to its geometry. The inference is the partial derivative that describes the rate of change of deformation

due to the change of geometry. For a constant force we deal with a smooth continuous function of deformation that has a limit. The derivative of such a function is the tangent to the curve in the interval of stable physical relations, namely the tangent in the interval $0 < \tan \alpha < 4$. The basic function should be in the same domain. This condition allows us to select an independent variable x for which $0 < dy/dx < 4$.

(8) The next task is the validation of the logical system of equations. The empirical validation of the system can be achieved by experimentally testing the basic equation. The logical justification of the system is achieved by establishing the correspondence of the basic equation to a geometric function that is known to be consistent and correct in the domain of physical stability, represented by the tangent function.

For building such system we select an independent variable concept that places the propositional function in the two intervals of the tangent function. In order to satisfy this condition an independent variable is selected that may have the same level of magnitude (10^b) as the deformation. The partial derivative equation from the basic equation (in case of static force) is the tangent function in the interval of rapid and slow changes. The character of changes described with this derivative function reflects the possible character of the physical changes in the individual structure. The character of the function shows the existence of a limit that depends on the geometry of the structure. The limit is characterized by a rapid increase of deformation. This means that the function of deformation should exist both in the interval of slow changes and in the interval of rapid changes.

The selection of a universal mathematical function, i.e. the tangent function, that describes similar mathematical relations and behavior gives a measure of logical agreement for the system consisting of derivative function and the corresponding basic function. It also gives unique certainty for the derivative values and corresponding values of the function in the basic equation. It was our purpose to build a logical system that allows us to reach certainty in a physical theory.

The correspondence of the basic and derivative functions to each other should be tested. This test can be done by comparing the empirically obtained deformation data with the theoretical curve. If necessary, the variables can be adjusted.

The experimental verification of the basic function gives empirical validation both for the basic function and for its derivative. However the independent validation can be obtained for the limit of elasticity in the interval of rapid changes.

(9) Because the non-linear theory of elasticity was built and the terms, mathematical statements and system of statements were logically justified we may assume that the rules of definitive logic can be useful in general. This logical procedure gives necessary and sufficient logical justification for a two-level mathematical structure that connects selected data with a conclusion. The conclusion is logically consistent for we select it identical to a mathematical-geometrical pattern that is known to be consistent and free of self-contradiction, in this particular case the tangent function. The conclusion is also in agreement with the character of physical changes of the phenomenon. The basic function is uniquely determined in each of its points for these points have known derivatives. The basic function can be tested experimentally for each of its points. The logical system and logical procedure described here have the possibility to provide and ensure a certain single solution while logically connecting the data with the conclusion.

This logical system is at the foundation of the non-linear theory of elasticity. I believe that definitive logic can be used in other areas of scientific inquiry, providing the same level of certainty to the logical connection of theory to phenomenon.

(10) There are mathematical procedures and there are logical connections between the parts of a theory. This does not mean that mathematics is identical to logic or vice versa. Logic provides justification for certain mathematical procedures, identifying the result as true or false. Falsification of a theory is rather connected with the empirical test of a proposition. We may say that for all practical purposes there is no equivalence between logic and mathematics.

The point of view here is that in order to be successful a physical theory needs a proven logical structure at its foundation. The logical structure in a physical theory attains mathematical forms and concrete physical substance and meaning. We discussed already how this system was implemented in the Non-Linear Theory of Elasticity.

Philosophers of science until now did not build logic for science, but rather analyzed scientific theories in order to find a common philosophy and logic in the theories. Although a common philosophy exists in the sciences, i.e. atomistic philosophy, there is no common logic except for deductive mathematical procedures. From the Afterword by Frederick Suppe to the second edition (1977) of *The Structure of Scientific Theories*: "Thus, at present philosophers of science still are searching for an analysis of theories which will provide an adequate philosophical understanding of theories." The main approach and purpose that F. Suppe indicates is "To what extent must an adequate analysis of scientific theories and scientific theorizing be an accurate description of what actually has transpired in the history of science or is happening in contemporary science." The task of establishing or recovering logical structure in scientific theories is not stated. In this work the author proposes a Logical Structure of Scientific Theory that should help in building a theory.

5.1 Recapture

Building a theory is a long historical process. Observation is the first step in the development of knowledge of a particular phenomenon. Observation of the elastic behavior of bodies started with the natural contact and adjustment of man to the surrounding world since its beginning. The technology of building things is often ahead of the understanding the physical essence of a phenomenon or the scientific presentation of this understanding. This first period is not a science yet.

Science starts with the formulation of some general idea concerning a phenomenon. For the theory of elasticity this generalization took place in 1675 when the English physicist Robert Hooke postulated that all solid bodies and all materials possess springing (elastic) properties. It is the power of word that starts science.

The phenomenon must be singled out and the effect (what?) is separated from the cause (why?). The description of an effect refers to an object in which the cause is absorbed as an effect. We do not include cause in this description. A function describes the relations among the selected appropriate characteristics of the effect or, in other words, of the phenomenon.

The main terms of a description, which are the physical characteristics of the phenomenon, need to be separated from each other. The study of the relations between those concepts then follows. In this stage, mistakes may arise that could be conditioned by the prevailing philosophy in science at the time of conception of the hypothesis and by the stage of technological development. It refers both to the selection of terms and to the assumed relations between them.

In this work the selection of terms in the non-linear theory of elasticity has been made as the result of a most general division. The terms are: the change of geometry of a body due to an external force, the internal elastic force arising in a body, the resistance of a body to deformation due to its geometry, and a coefficient – the modulus of elasticity of the material. The selected terms and the function that is constructed to describe the relations between the concepts are then tested experimentally.

An experiment is usually designed to confirm a hypothesis rather than to falsify it. However, the experiments may support or disprove our assumptions. The empirical support of a function, though important, does not yet prove a theory, no matter how many experiments are performed.

Proof can only be reached logically, within some logical system that allows a mathematical deductive conclusion and an inductive conclusion from a general law to the unknown particular laws. This is the nature of logical truth. The logical system of the theory of elasticity consists of a basic equation, describing the relations among elastic force, geometrical stiffness of a body and change of geometry of a body, and a derivative equation showing the rate of change of the basic function. The basic function and the derivative function are in a necessary logical–mathematical conjunction.

Geometrically, each point of the curve representing the derivative function is the value of the tangent to the curve representing the basic function. Analytically it is the ratio of the unit of change of the basic function at some point to the unit of the independent variable. Although each point of the derivative function can be found as the tangent to the curve representing basic function, in general the derivative function can be any function. In the non-linear theory of elasticity, for a constant elastic force this derivative happens to be the tangent function. In case of a dynamic force acting on a body the deformation would follow the character of force. The derivative equation will correspond to the deformation. But in all cases the domain of possible existence of the physical phenomenon of elasticity is in the interval of the tangent function $0 < \tan\alpha < 4$. Seemingly an abstract mathematical relation indicates the domain of possibility and impossibility of physical existence.

The difference between describing the rate of change of a hypothetical function that is not logically formulated and the rate described with a universal function that is proven to be correct is that an unproved unstructured derivative leaves the logical system yet unproved. On the other hand, a derivative function that is known to be mathematically consistent makes the system logically and mathematically correct in the domain of possible physical existence.

We deal with functions describing a physical phenomenon. The question is, can we choose a logical system without violating physical correctness. The proof of physical correctness of the system comes in two ways, that is, empirical verification of the basic function and correspondence of the basic function to a known universal derivative function. The derivative in turn should satisfy the condition of possibility of physical existence of the basic function in question. Thus the phenomenon of elasticity demonstrates its existence in the interval of slow linear changes of deformation, in the interval of rapid changes, and in a smooth transitional interval. For a static force the equation of deformation is a smooth continuous parabolic function. The derivative that satisfies this description and contains the tangents of all of its points is the tangent function. Although we have no empirical method for confirming the tangent for each point of a function, the fact that we empirically confirmed the corresponding basic function shows that the system gives the necessary conjunction to the derivative function and its physical correctness. If the derivative is completely mathematically certain then the corresponding basic equation possesses the same mathematical certainty. On the other hand, the derivative of a hypothetical basic function might also be a hypothetical function. Experimental support for the basic function is not sufficient for confirming the hypothesis. We need confirmation of the inference. The relative character of the derivative function prevents independent empirical verification. The derivative function is logical–mathematical in nature and requires logical–mathematical proof.

The definitive logical–mathematical system describing a physical phenomenon does not leave much to imagination. The construction of such a system is a rather mechanistic job. An example is presented in Flow Chart at the end of section 7.5.

The scientist needs imagination and mental experimentation for the explanation of physical data that is not included in the mathematical description and not covered by some universal law. The explanation can be or can seem to be logical, but it is nevertheless hypothetical even if supported empirically. The hypothesis that the limit of elasticity of a structure can be of different origins is supported both logically and experimentally.

The calculation of an individual limit of elasticity for a structure finds its support in experiments on specimens with different geometrical stiffnesses. It is existentially correct until the calculated limit of elasticity exceeds the limit of the material. Beyond that, elastic failure of the specimen occurs when the stress reaches the limit of elasticity of the material. My explanation was that we could divide a body at different levels of structural uniformity. Each level has its own limiting characteristics. The real limit that reveals itself in a destructive experiment on a structure is the relatively smallest value. This explanation is logical because the logical–mathematical system shows the existence of a system-specific limit. Materials tests show the existence of a limit depending on the inherent elastic properties of a material. Mathematical logic suggests that when the elastic force reaches the lesser of these two limits the structure reaches the limit of the relation that is described by Hooke's law, $\sigma/\varepsilon = E$.

Hooke's law is covered by inductive inference from the general law of conservation of energy. Therefore it is logically justified. Empirically this law is justified by testing material specimens. The comparative analysis of data on limits is also mathematically and empirically justified. It is quite possible that when an experimenter in the observation of, or experiment on, some physical phenomenon encounters two different manifestations of the phenomenon, it may indicate different levels of origin of the characteristics. In the theory of elasticity, the limiting characteristics originate in the material or in the geometry.

6. It is possible to prove physical theory

We have concluded that the logical structure of a physical theory needs to be proven and can be proven. However, this point of view is not yet common. Building a theory on a logical foundation is not the common approach in science. In the literature we rather arrive at a different position. The common belief is that there is no method for constructing and proving a physical theory. Here follows an expression of the almost unanimous current position: "The laws of logic and pure mathematics, by their very nature, cannot be used as a basis for scientific explanation because they tell us nothing that distinguishes the actual world from some other possible world. When we ask for the explanation of a fact, a particular observation in the actual world, we must make use of empirical laws. They do not possess the certainty of logical and mathematical laws, but they do tell us something about the structure of the world." (Rudolf Carnap, An Introduction to the Philosophy of Science, edited by Martin Gardner).

The scientist of contemporary science separates logic from physical observations, from the building of a physical theory, and from the test of a physical theory. "In the field of the empirical sciences a scientist, whether theorist or experimenter, puts forward statements or a system of statements, and tests them step by step by observation and experiments." (The Logic of Scientific Discovery, Karl R. Popper). There is no reference to some logical structure being necessary for constructing and testing a theory. This renowned philosopher, whose viewpoints are still influential, sees the only way to support a theory in the observations and experiments. However it is widely accepted that empirical methods cannot prove a theory. Based on this fact the generalization was made that a physical theory cannot be proved, but it can be disproved.

"The fact that a law has been substantiated in $n-1$ cases affords no logical guarantee that it will be substantiated in the nth case also, no matter how large we take n to be. And this, we shall find, applies not only to general propositions but to all propositions which have a factual content." (Language, Truth and Logic, Alfred Jules Ayer). The problem is not in the possibility that the sun disappears from the sky on day $n+1$ but in the fact that, as Alfred Ayer says, "They [factual propositions] can none of them ever become logically certain."

Unless we construct our hypothetical propositions according to the rules of logic we cannot attain logical certainty of the theory. The attitude to the role of logic in science prevents reaching logical certainty in the scientific theories. In "An Introduction to the philosophy of Science" by Rudolf Carnap, edited by Martin Gardner, we read: "The observations we make in everyday life as well as the more systematic observations of science reveal certain repetitions or regularities in the world. The laws of science are nothing more than statements expressing these regularities as precisely as possible." Observations may reveal but mostly do not reveal regularities that can be considered as laws. And although observation clearly reveals the regularity that solid bodies have springing qualities, many millennia had passed before Robert Hooke announced it as the law of proportional relation between force and deformation.

R. Carnap's next assertion, "statements of logic and mathematics do not tell us anything about the world" is an almost generally accepted point of view. The statements of logic and mathematics tell us nothing because we ask them nothing. In physics, logical and mathematical statements acquire physical substance and thus do tell us something about the world.

The validation of a scientific law by both empirical and logical means is another question. The answer depends on the position we come from. If it is the position that for the validation of a hypothesis we need as many facts as possible, then we will never prove the hypothesis.

Thus, the equation $e = NL/EA$ of elastic elongation cannot be proven, no matter how many facts, and they are innumerable, show that the equation gives satisfactory results for predicting the elongation of a simple bar. This equation cannot be proven because it is not included in any logical structure, and that is the only way to prove a hypothetical statement. As soon as we try to construct a logical–mathematical structure based on this equation it becomes apparent that the equation is improper. It does not describe the relations between the variables of equation correctly. The logical system applied to the physical theory changes the explanation of the physical essence of the theory of elasticity, and tells us not less but rather more than the physical facts that have been collected in centuries of intensive practice.

If, on the other hand, we approach the building of a theory from the point of view of logical necessity then we may expect a theory that can be proven. For example, we may have logical certainty of a statement that concerns physical matter: If a physical relation has a limit then this relation requires a description with a non-linear function. The backward assertion is also correct, namely, if a physical relation must be described with a non-linear function, then this relation has a limit depending on the variables in this function. This logical conclusion concerns physical facts and it is reached by inductive logic because we know the character of the mathematical description. Mathematics in physics tells us something about physics. The logical structure consisting of basic function and derivative allows us to predict this limit.

A propositional statement should be in conformation not only with physical but also with logical demands. In order to describe the rate of deformation correctly the function of deformation should occupy the domain of the tangent function both in the interval of slow changes and in the interval of rapid changes. This logical demand is connected with the knowledge of the law of mathematical deduction for the rate of a function and the inductive conclusion based on the knowledge of the tangent function that is associated with the description of the rate.

A logical conclusion concerning physical relations is that in order to prove physical propositions we need a logical system consisting of basic and derivative functions. The principle of uniqueness of a physical function requires these two interconnected functions to describe both the absolute values of the function corresponding to a specific structure and the position of each structure in a set of similar structures. We need proof for a system rather than empirical verification of the hypothetical basic equation, for such verification by itself has no certainty and no logical proof. However, if the necessity of the particular logical structure is proven then each part of the structure has logical certainty. The empirical verification of a basic function gives at the same time such verification for the derivative values.

It is a logical conclusion that a derivative function that has a mathematical–logical property as its result can be substantiated only by identifying it with the geometrical logical structure. "All that the geometry itself tells us is that if anything can be brought under the definitions, it will also satisfy the theorems. It is therefore a purely logical system, and its propositions are purely analytic propositions." (Language, Truth and Logic, Alfred J. Ayer). And further on the logical nature of geometry: "There is no sense, therefore, in asking which of the various geometries known to us are false and which are true. In so far as they are all free from contradiction, they are all true." (A.J. Ayer). The physical basic equation that is supported with facts and included in the logical system also satisfies the demand of logical certainty.

On the other hand, separation of logic from physical observations, lack of clarity on the roles of logic, mathematics and physics in physical theory, and mixed identities of logic and mathematics bring uncertainty to a scientific theory. Not all philosophers agree that logic has no part in the explanation of physical reality. Morris R. Cohen in his "A Preface to Logic" says, "Reflection shows that logic cannot be isolated from any realm of being ... " and "The assumption that the objects of physics and other sciences must conform to logic is necessary in the sense that without it no science at all can be constructed." (M.R. Cohen).

7. Notes on logic

7.1 Commentaries to "Preface to Logic" by Morris R. Cohen

> Nearly all the books define logic as in some manner the science of thought. But that the laws of logic are not the universal laws according to which we do actually think is conclusively shown, not only by the most elementary observation or introspection, but by the very existence of fallacies. *(M.R. Cohen)*

The laws of logic are not the laws on how we think irrespectively of our fallacies or of logic fallacies. Speaking of logic fallacies we have to notice that the basic argument of classical logic accepts as equal possibilities that a conclusion is true if the premises are true, and a conclusion is false if the premises are not true. The laws of logic are not about how we really think or how we have to think. We think without consulting books on logic. Knowledge of the laws of logic no more improves our ability to think than does any other systemic knowledge. Logic and its laws are not constructed according to the thought-train even of the smartest of human beings. In science we have to consider logic as a scientific tool for establishing the necessary connections among things or properties and for justifying these connections. A different point of view prevails in the current philosophy of science.

"Logic is the most general of all the sciences; it deals with the elements or operations common to all of them. That is, rules of logic are the rules of operation or transformation according to which all possible objects can be combined." (M.R. Cohen). This definition of logic is not specific enough. The same definition can be applied to mathematics as "the rules of operation and transformation". Some of the laws of logic may be similar to the laws of mathematics. Logic concerns itself with the methods of making a proper reliable inference from an initial proposition. And calculus as well gives us methods for obtaining a derivative from the initial function. Often logic is identified with mathematics. "Logic and mathematics (the two, we shall see, are identical in essence) explore and examine the meaning or implication of any proposition, and no proposition about nature can pretend to scientific truth unless it submits to such an examination of its meaning or logical consequences." (L. Pearce Williams, The Origins of Field Theory). We should note that mathematics do not "explore and examine the meaning and implication of any proposition". That is not the purpose of pure mathematics.

"The nature of the subject matter of logic may be better understood when it is seen to be identical with the subject matter of pure mathematics. ... Algebraic proofs are in every respect logical proofs and depend no more on any special element of intuition than does logic itself." (M.R. Cohen). The logic that is identified with mathematics loses its own identity. In particular, this logic misses a part that makes it definitely distinguishable from all other sciences, that is, it misses the process of establishing proof of a hypothetical proposition. Including the proposition in a logical structure can do this. Logic, unlike pure mathematics, should be concerned not only with the formal truth of a system but also with its physical or material truth.

We have to apply logic in the process of selecting and evaluating terms for the proposition and the construction of the propositional function. This is not a task for mathematics. Some terms can be evaluated empirically, but this does not mean that they necessarily fit into the logical structure. As well, empirical validation of the propositional function is not yet logical truth and therefore cannot be accepted as proof of correctness of a physical function. Further, algebraic proof from assumption to implication is not a sufficient logical procedure for establishing the material truth of a conclusion.

The identification of logic or theoretical physics with mathematics leads to insufficiency for both logic and physics. No scientific discipline can be simply part of another discipline. Logic, mathematics and physics have specific distinctive purposes, methods and roles in physical theory. Logic provides the structure that can proof a theory. This structure is presented in the form of interconnected mathematical equations. Logic requires the selection of physical variables/concepts that can satisfy the strict logical structure. The selection of the variables should be logically justified. The construction of the mathematical statements and their connection needs logical justification. Thus, the main role of logic in physical theory is to provide justification for the parts of a theory and the theory as a whole.

Logic has two methods, inductive and deductive, to do the job. Morris R. Cohen wrote: “We have used the term logic or formal logic as identical with deduction.” And further “Deductive logic and pure mathematics generally deal with certain relations between propositions, and the knowledge of such relations is certainly one of the most potent instruments of scientific research.” We have already discussed that a more differentiated approach is needed between pure mathematics and the logic of deduction.

On the method of induction Morris R. Cohen wrote: “The term

induction has been used to denote among other things: 1. Reasoning from facts or particulars to laws or universals (Boethius and the scholastics). 2. Reasoning which is based on the principle of uniformity of nature, i.e., like effects must have like causes (Mill). 3. Disjunctive reasoning (Schuppe, Montegue)."

With the understanding that none of these approaches provides justification for a conclusion M.R. Cohen wrote: "If anyone thinks that I have understated the case for these canons of induction as methods of discovery, let him discover by their means the cause of cancer or of disorders in internal secretions."

More productively, the method of induction can be considered as making inferences from universal general laws to particular statements and terms. Although it is impossible to justify inference from particulars to a general conclusion, it is possible to justify inference from accepted general laws, theorems, and definitions to particulars.

Logic provides a frame for a physical theory. Morris R. Cohen eloquently expressed his point of view on this role of logic: "And it is as impossible to derive physical or psychological truth from pure logic as to build a house with nothing except the rules of architecture. The form or structure of a house is constituted by the system of relations between the material entities which make it up; and the form or structure which logic studies is the system of relations which hold between all possible objects that can be ordered in the system."

The point of view here is that logical truth is connected with physical truth. One of the purposes of logic, at least in this work, is constructing a frame for a physical theory. The task of constructing a physical theory requires separating the role of logic from that of mathematics. There is a principal difference between the laws of logic and the laws of mathematics. The purpose of the laws of logic is to get to the truth of the matter of facts. In mathematics we are not concerned with truth of matter of facts, for mathematics does not deal with matter. "So long as you are dealing with pure mathematics, you are in the realm of complete and absolute abstraction. All you assert is, that reason insists on the admission that, if any entities whatever have any relations which satisfy such-and-such purely abstract conditions, then they must have other relations which satisfy other purely abstract conditions." (N. Whitehead, Science and the

Modern World). The difference, as we can see, is in the essence, even though a logic procedure of inference includes a mathematical procedure, and quantitative physical relations are obtained by means of mathematical manipulations.

The comparison between the general purpose that we have in physics and the purpose of logic, on the other hand, shows a similarity. Both are concerned with establishing the truth of initially hypothetical relations that is seen and understood in the correspondence of inference to the physical facts. But because formal logic has no means to prove the truth of premises it distinguishes factual truth from formal truth. Then, "The distinctive subject matter of logic, constituting, as a matter of fact, the core of the traditional Aristotelian logic, is what is called formal truth." (M.R. Cohen).

The purpose of logical procedure, as presented in this work on the example of constructing the non-linear theory of elasticity, is to eliminate the difference between the factual truth and the formal truth. The truth reached by means of logical–mathematical consequences from a proposition can be proved to be factual truth. Often our initial physical knowledge that refers to observation the results of physical interactions is accepted as a factual truth. The characteristics of a phenomenon are selected subsequently in order to obtain the independently known result of the interactions. Thus, in the non-linear theory of elasticity, by analyzing the possible result of structural elastic behavior depending on the rate of change of deformation we arrive at the conclusion that this behavior occurs in the domain of two intervals of the tangent function, and the inference can be identified with the tangent function. The tangent function in this contest is a logical structure that reflects physical behavior as well. In the interval of rapid changes we may observe elastic failure and destruction of a structure. In the interval of slow changes the behavior of a structure is predictably elastic. The logical conclusion is that we have to describe the initial conditions in such a way that inference from this description provides the observed result. The function we construct as a basic description should occupy the domain of the tangent function in both its intervals. The mathematical description of the basic equation and its known derivative is the practical realization of the necessary logical structure. This logical construction at the foundation of the theory was made not in accordance with the rules of formal logic but rather with the rules of definitive logic that allows flexibility. We may exchange the places of assumption and conclusion and make a description starting with the conclusion as known from observation, then progressing to the necessary proposition. Every single step in such construction of a theory is based on logical justification, and thus we arrive at a system that can be verified. We do not neglect facts. But the selection of facts follows the logical necessity. The temporal physical connection, what is first and what is consequent, does not place a restriction on our reasoning. For building a reliable theory we have to go backward from the known result to the unknown conditions. Philosophers and logicians distinguish 'a priori' knowledge from 'empirical' knowledge.

"Logical truth or consistency is a genuine part of the world of truth which science studies." (M.R. Cohen).

7.2 Commentaries to "An Introduction to the Philosophy of Science" by Rudolf Carnap

7.2.1 Definition of scientific law

The definition of scientific law according to Carnap is that "The observations we make in everyday life as well as more systematic observations of science reveal certain repetitions or regularities in the world. The laws of science are nothing more than statements expressing these regularities as precisely as possible."

From our point of view in this book no amount of observations, even if they have led to some general idea, brings us to the understanding of relations in nature. The observations do not reflect objectively the relations between the characteristics of a phenomenon. The terms participating in description of a phenomenon may be observable while others are non-observable. Complex work involving both observations and logical procedure is needed for formulating the laws of science.

Carnap divided scientific laws into 'universal' and 'statistical'. If a certain regularity is observed, then the regularity is expressed in the form of universal law. Universal laws are stated in the form of formal logic as a universal conditional statement. The simplest general assertion is that, whatever x may be, if x is P, then x is also Q. This is written symbolically as follows: $(x)(\mathrm{P}x \supset \mathrm{Q}x)$. The (x) is called the 'universal quantifier'. Px means x is P, and Qx means that x is Q. The symbol $\supset$ is connective. In English it is the assertion "If P, then Q". The general schema involved in all explanations of the deductive variety can be expressed symbolically as follows: 1. $(x)(\mathrm{P}x \supset \mathrm{Q}x)$; 2. P$a$; 3. Q$a$. The first and second statements together enable us to logically derive the third statement. "It is true that the laws of logic and pure mathematics are universal, but they tell us nothing about the world." (R. Carnap) I should say that the laws of logic tell us nothing about the physical world unless we apply them to specific facts, for instance to a physical phenomenon. Then these laws attain physical essence. Carnap says "simple logical laws such as 1. If p and q, then p. 2. If p, then p or q. Those statements cannot be contested because their truth is based on the meanings of the terms involved." These statements, as I see them, include also the assumption that, at least, assumption P can be tested; and also the assumption that there is no need for an independent test of assertion Q. There is a somewhat different assumption in modern deductive logic. "An argument is defined as a valid deductive argument if its premises could not all be true while its conclusion was false." (Robert J. Ackerman, Modern Deductive Logic). This definition assumes that the conclusion can be tested empirically, which is not always the case: in physics it is often a pure mathematical–logical property that cannot be tested empirically.

The logical structure of the non-linear theory of elasticity and the whole class of physical theories has inductive assumptions and deductive conclusions. The current laws in logic are insufficient for proving the correctness or falsity of each statement, and tell us nothing of the factual relation between P and Q. P is a scientific proposition that is based on observation and experiments. Q is a mathematical inference that necessarily follows from P. But this logical structure is insufficient for describing the logical relations between P and Q. The description should include the necessary inductive procedures for establishing a connection between statements P and Q and observation. For example, the current theory of elasticity has a mathematical statement that is consistent with the observations from the infinite number of tests. The deformation of a bar is proportional to the tensile force applied to the bar, proportional to the length of the bar, and inversely proportional to the area of cross-section and the modulus of elasticity of material. The mathematical conclusion that can be drawn from this equation is a partial differential equation that is true mathematically and is assumed to be logically true by necessity. However, from this procedure we are unable to establish definitely that it is true or false. Without independent logical proof of the consistency of the conclusion we have no logical proof for our empirically obtained proposition. The conclusion Q is a formal logical conclusion that implies the assumption that we described the facts correctly. There are many ways to describe the relations between the same set of the facts. But, contrary to the prevalent point of view, only one of all these possible descriptions may give the conclusion that can be proved as true logically and empirically.

We cannot prove statement Q empirically by itself, for it is mathematical in nature. Nonetheless, the character of the relations it describes should be consistent with the physical nature of the phenomenon that we describe. As with the construction of the statement P, it is inductive logic that helps to establish the empirical validity of the statement Q. By itself no inductive procedure accounts for the logical proof of our construction. If Q is inconsistent with observation we should select another description of Q that is consistent with fact and then construct the corresponding statement P. Such new P gives the same result as the previous P and it is presently considered to be an equivalent description. However, one cannot have equivalent descriptions for the same physical phenomenon. The mathematical inferences from the different mathematical statements are different. The physical implications from such changed statements are quite different.

When changes to P have been made and we reach the necessary conclusion that is consistent with the observation of the phenomenon we still cannot say that our logical–mathematical structure is correct, for Q still has no independent proof of its existential validity. No amount of empirical validation provides proof of the logical correctness of a system of statements. The mathematical conclusion Q describes the relations and has a mathematical property as its result that cannot be validated independently without an assumption. Nonetheless, if the relation it describes is consistent with the observations we may find an abstract model of such relations in mathematical geometry that provides the logical model. In our particular case such model is provided by the tangent function. This function gives logical consistency to the values of Q in the complete domain of function P. It makes, on the one hand, function P logically consistent and definitive, and the derivative Q definitive as well by the definition of the differential procedure. On the other hand, the mathematical conjunction of P and Q provides empirical validation to the abstractly obtained values of statement Q.

In many cases the law may be statistical. The predictions then will be probabilities only. In view of our discussion, Carnap's conclusion "When the law is universal, the elementary deductive logic is involved in inferring unknown facts. If the law is statistical, we must use a different logic – the logic of probability." is obviously a simplification, but it is also methodologically incorrect. Let us consider why.

7.2.2 Induction

Rudolf Carnap distinguishes two types of logic: inductive and deductive. All laws, in his opinion, are based on the observation of certain regularities. What justifies us in going from the direct observation of facts to a law? This is recognized as 'the problem of induction'. "In deductive logic, inference leads from a set of premises to a conclusion just as certain as the premises. If the premises are true, the conclusion cannot be false. With respect to induction, the situation is entirely different. The truth of an inductive conclusion is never certain. Even if the premises are assumed to be true and the inference is a valid inductive inference, the conclusion may be false. The most we can say is that, with respect to given premises, the conclusion has a certain degree of probability." (R. Carnap).

The point of view in the definitive logic is that premises are mainly of inductive character. The initial proposition is hypothetical in nature not because some of its terms have been obtained with the inductive process but because it can be proven only in the logical conjunction with inference. The inference has the same degree of uncertainty as the premises. The following conclusion in classical logic is based on the recognition of this fact. "At no time is it possible to arrive at complete verification of a law but only of confirmation." (R. Carnap). "Although there is no way in which a law can be verified (In the strict sense), there is a simple way it can be falsified. One need find only a single counterinstance." (R. Carnap).

In fact, strict logical verification, as was shown on the example of the non-linear theory of elasticity, is possible by identification of the derivative conclusion with a geometrical function describing similar logical relations. Falsification of a theory or part of a theory, on the other hand, should not be so easily accepted. If we are confronted with two facts: on the one hand the observations of a phenomenon are correct and provide a description that is proved time after time, but the mathematical conclusion from this description is false, then we may think of selecting different, probably compound terms of description. We cannot disregard the facts, but we can change the description by selecting different characteristics. The selection of the characteristics of a phenomenon is an inductive process, in which truth is obtained with the process described before and presented in definitive logic.

In deductive logic we go from a proposition to the necessary inference. In the traditional logic, if the premises are true, the conclusion cannot be false. In modern deductive logic we deal with the negation of the conclusion. If the conclusion is wrong, then the assumptions cannot be all correct. In the definitive logic described in this work, if the premises are empirically supported and the inference is logically proven to be correct (self-consistent) and consistent in character with the observation in the domain of possible stable physical existence, then the premises are logically correct as well. This system is uniquely verifiable.

Confirmation of physical theories or laws in classical logic refers to physical tests and observations only. "How do we find confirmation of a law? If we have observed a great many positive instances and no negative instance, we say that the confirmation is strong. There are, of course, various methodological rules for efficient testing." (R. Carnap).

We already discussed that confirmation cannot be based on empirical methods only. A provable logical structure should be introduced into a theory. Instead, current logic often uses the concept of probability of a law. Carnap distinguishes two types of probabilities: "It is my belief, however, that there are two fundamentally different kinds of probability, and I distinguish between them by calling one 'statistical probability', and the other 'logical probability'. Instead of 'logical probability' I sometimes use the term 'inductive probability'. ... By 'inductive inference' I mean not only inference from facts to laws, but also any inference that is 'non-demonstrative'; that is, an inference such that the conclusion does not follow with logical necessity when the truth of the premises is granted. Such inferences must be expressed in degrees of what I call 'logical probability' or 'inductive probability'." (R. Carnap).

There is no physical science that could demonstrate a conclusion that is made from the facts to the general proposition. The opposite process is more likely: having a proposition we look for facts that support the proposition. The idea in physics appears first, as a rule, in connection with other similar ideas and then it is connected to the facts. "Traditional logic always leads to the solution with a certain degree of probability. Though the inductive procedure of selecting physical concept-characteristics is connected with some degree of probability, but this uncertainty disappears in further logical development of a theory." (R. Carnap).

Probability, or rather accuracy of the results that depends on measurement is always present. Thus, the probability of the compound coefficient of specifics K in the NLTE is a statistical probability. We cannot assure that the geometrical stiffnesses of similar structures have absolutely the same value for K. The correctness of such an approximation can be statistical only. A probability of errors in obtaining the non-observable and to some extent hypothetical characteristics of a phenomenon also exists. Thus, stress in the theory of elasticity is such unobservable characteristic of a deformed body. The existence of stress is an inductive conclusion that is based on the fact that deformation disappears after the external forces applied to a body are removed. In our system of knowledge we connect the definition of work with the necessity of there being a force that performs the work. Although Carnap thinks that it is possible to apply an inductive logic to the language of science, he continues however "I do not mean that it is possible to formulate a set of rules that will lead automatically from facts to theories."

Inductive rules can probably be formulated for the different classes of facts, but inductive logic by itself as well as deductive logic by itself cannot lead to a definitive conclusion. A logical system that has parts justified with the inductive method as well as parts obtained with the deductive method is more likely to provide a reliable physical law. Concerning Carnap's point "that there cannot be an inductive machine – a computer into which we can put all the relevant observational sentences and get, as an output, a neat system of laws that will explain the observed phenomena.", I believe that, while some explanations would be left for our imagination, a comprehensive and reliable set of laws can be formulated using the rules of inductive and deductive logic. Carnap proposes: "The main points that I wish to stress here are these: Both types of probability – statistical and logical – may occur together in the same chain of reasoning. To statements about statistical probability we can apply to logical probability, which is part of the metalanguage of science."

In the productive process of building a physical theory I do not see a place for logical probability for it would interfere with the purpose of definiteness of logic in a physical theory.

7.2.3 Concepts in science

Carnap says that concepts in science can be divided into three groups: classificatory, comparative, and quantitative. "It often happens that a comparative concept later becomes the basis for a quantitative one" (R. Carnap). Comparative concepts imply a complicated structure of logical relations. We see that there are two ways in which the comparative concepts of science are not entirely conventional: They must apply to facts of nature, and they must conform to a logical structure of relations.

The difference between qualitative and quantitative concept that Carnap proposes is a difference in language. "The qualitative language is restricted to predicates (for example, 'grass is green'), while the quantitative language introduces what are called functor symbols, that is symbols for functions that have numerical values. ... Some philosophers maintain that modern science, because it restricts its attention more and more to quantitative features, neglects the qualitative aspects of nature and so gives an entirely distorted picture of the world." (R. Carnap).

I think that it is worth mentioning that qualitative features are connected with quantitative values. This is true for concepts and can be seen in a set of similar concepts. Thus, the ability to resist changes depends on the value of geometrical stiffness. In a diagram of deformation vs. geometrical stiffness of otherwise similar structures we see how quantitative changes of geometrical stiffness in one interval do not lead to qualitative changes but in another interval lead to a drastic change of elastic behavior. Attention should be paid to this fact that the quality of some properties can change with a change in their magnitude. This effect should be recognized and evaluate, which is not always done in quantitative analysis.

7.2.4 Measurement

"We will now take up a question that has been raised many times by philosophers: can measurements be applied to every aspect of nature? Is it possible that certain aspects of the world, or even certain kinds of phenomena, are in principle non-measurable?" Carnap's point of view is that "It is *we* who assign numbers to nature. The phenomena themselves exhibit only qualities that we observe. Every thing numerical, except for the cardinal numbers that can be correlated with discrete objects, is brought in by ourselves when we devise procedures for measurement." While we devise the procedures, this does not mean that measurement is an invention without objective basis. It is also does not mean that the magnitudes we assign to concepts are always objective values. The values we assign to compound concepts that are not directly observable often include our assumptions. For example, the magnitude of stress, which we cannot measure directly, depends on our assumption of the stress distribution and on the method of calculating this magnitude, for stress by itself is not a directly observable concept and cannot be measured directly. What we can measure is deformation or external force and the dimensions of a structure. The magnitude of stress depends on what assumption is adopted. Carnap states: "When we assign numbers to phenomena, there is no point in asking whether they are the 'right' numbers. We simply devise rules that specify how the numbers are to be assigned. From this point of view, nothing is in principle unmeasurable."

Although numbers can be assigned and are assigned to physical concepts, we still should ask the question "Are they the right numbers?", for it is important to get correct physical results from the law. If the magnitude of stress is calculated with the assumption that the stress distribution follows the distribution of external forces, then the magnitude of stress might be twice that calculated with the assumption according to Hooke's law that stresses are distributed evenly in the material of a structure. Our assumption in this case does obviously matter. From the knowledge of stress distribution we derive the size of a structure and the amount of material we use. We cannot arbitrarily assign magnitudes to concepts that are not measurable directly, but we should rather provide empirical and inductive (explanable) support for these magnitudes, or, if possible, get rid of them during measurement and then compare the values obtained with direct measurement and with using conceptual magnitudes. For example, the calculation of optimal geometrical stiffness can be done without knowledge of stress from the system of equations $e = N/ER_a$, $N/ER_o^2 = C_s$. From this system of equations, by measuring the deformation e in a prototype of the structure and measuring the actual geometrical stiffness of the prototype R_a we can correct the geometrical stiffness of the structure, $R_o = \sqrt{eR_a/C_s}$. Knowledge of the law and measurement of the observable characteristics allow us to get an optimized result while avoiding a non-observable concept. This knowledge also allows us to obtain the magnitude of the non-observable theoretical concept of force, $N = eER_a$, and to compare it with the values of distributed forces according to the assumption of static equilibrium of external and internal forces for any part of a structure, which is presently a working assumption. From our point of view in this book this assumption is wrong.

"If there are a dozen different ways to measure a certain physical magnitude, such as length, then instead of a single concept of length, should we not speak of a dozen different concepts?" Carnap cited the view of philosopher P.W. Bridgman from his work "the Logic of Modern Physics". This is the view that every quantitative concept must be defined by the rules involved in the procedure of measuring it. It is not of course the point of view in my work. A definitive analysis of the concepts involved in the law should be carried out in order to obtain the most logically consistent concepts. The measurement is connected with the law. "It seems best to adapt the language form used by most physicists and regard length, mass, and so on as theoretical concepts rather than observational concepts explicitly defined by certain procedures of measurement." (R. Carnap). This is not what is defined as theoretical concepts in this work. What about the rate of change of a function? Is it a measurable or non-measurable concept?

Carnap says: "Quantitative concepts are not given by nature; they arise from our practice of applying numbers to natural phenomena." And further "The most important advantage of quantitative law, however, is not its brevity, but rather the use that can be made of it. Once we have law in numerical form, we can employ that powerful part of deductive logic we call mathematics and, in that way, make predictions." Quantitative concepts are objective concepts. And it is not mathematics that makes predictions, but rather the correct logical structure that allows us to make predictions.

7.2.5 Geometry and a theory

> The nature of geometry in physics is a topic of great importance in the philosophy of science. ... First of all it leads to the analysis of the space–time system, the basic structure of modern physics. Moreover, mathematical geometry and physical geometry are excellent paradigms of two fundamentally different ways of gaining knowledge: the aprioristic and the empirical. If we clearly understand the distinction between these two geometries, we shall obtain valuable insights into important methodological problems in the theory of knowledge. *(R. Carnap)*

We already considered different applications of geometry to the equation of deformation. This equation is used for solving problems concerning a physical body, for instance to find the deflection of a beam. The differential equation for the elastic line is used to find the angular deformation of a cross-section. On the other hand, the equation of deformation in the framework of constructing the logical structure of the theory of elasticity is used for the logical comparative analysis in a set of similar structures. The curve we obtain in this case is a mathematical function rather than the description of a physical body. The derivative differential equation in this case is a mathematical function analogous to the tangent function. The system of these two functions is a logical system. The validity of this system for the specific phenomenon of elasticity is not only in the observational validity of the basic equation that is based on the particular method of observation and subjectively selected observable properties, but also in the objective validity of the derivative function. Conceiving a law is basically an inductive process, but this process can be objective. The objective character of physical relations is supported by the means of deductive logic.

Currently this combination of logical procedures has been cut short and the methodology has become too complex and uncertain to follow up. "A method for distinguishing between a purely logical derivation and a derivation that brings in non-logical components based on intuition became available only after the development of a systematized logic in the second half of the last century [i.e. 19th century]. The fact that this new logic was formulated in symbols increased its efficiency, but it was not absolutely essential." (R. Carnap). This is essentially the same logical argument as in the formal classical logic: If one believes in the argument, one believes in the conclusion. My system is not based on beliefs.

The current point of view is that negative statements asserting impossibility are much harder to prove than positive statements. The positive statement that this or that can be derived from certain premises is demonstrated simply by showing the logical steps of derivation. Our point of view in this work is rather the opposite, namely that a positive derivative statement from the premises does not yet affirm the correctness of the system of functions. On the other hand, a negative empirical test of propositional function does prove the impossibility of the proposition. "There is an infinity of different ways that physicist could describe their world, and, according to Poincaré, it is entirely a matter of convention which way they choose." (R. Carnap). Our point of view in this book is that there is determinism in a physical phenomenon, and accordingly our way to describe the phenomenon should be deterministic as well and depend on the rules.

"At this point, it is well to distinguish clearly between what we mean here by equivalent theories and what is sometimes meant by this phrase. Occasionally two physicists will propose two different theories to account for the same set of facts. Both theories may successfully explain this set of facts, but the theories may not be the same with respect to observations not yet made. That is, they may contain different predictions about what may be observed at some future time. Even though two such theories account completely for known observations, they should be regarded as essentially different physical theories." (R. Carnap). And further "It is very important to understand that 'equivalent theories', as used here, means something much stronger than the fact that two theories account for all known observations. Equivalence here means that two theories lead in all cases to exactly the same predictions."

It is quite obvious that the non-linear theory of elasticity is not equivalent to the linear theory of elasticity. As a matter of fact these two theories lead to different predictions for the limit of elasticity of a structure and to different presentations of elastic behavior, though both theories may use the same initial data. Aside from that, the linear theory uses an arbitrarily selected factor of safety and correlative coefficients for each of the variables in the function. Those correlative coefficients are not related to the limit of elasticity of the material. There is no sense in applying them to establish the allowable stress based on the limit of material. The only way to predict the individual limit of elasticity of a structure with the linear theory of elasticity is a destructive experiment on that particular structure. There is also no criterion for the similarity of structures, so that the destruction of one structure does not guarantee that similar structures different only in size can be safely built based on that experiment. The non-linear theory of elasticity operates with a general universal criterion describing the rate of change of a function. This new criterion is the tangent in the interval of significant figures, more accurately in the interval $0 < \tan \alpha < 4$. There is, probably, genuine misunderstanding about what designates equivalence of descriptions of physical phenomena. Thus, despite clear physical and logical differences in the linear and non-linear theories of elasticity, reviewers of the latter, presumably scientists from the [American] National Science Foundation, responded that "It [non-linear theory] is not new. Such tests are available." These tests are usually designed to support or reject our assumptions. It is impossible only by means of empirical tests to attain an objective judgment concerning physical theory. Only the conjunction of logic and empirical validation provides proof for a theory. I do not believe that two theories in the same system of knowledge can be equivalent. In order to obtain the same results based on the same set of facts the logical structures of the theories should be the same. There are no equivalent physical theories. The linear and non-linear theories, even though referring to the same data, are not equivalent.

The three geometries referred to by Carnap are applied to different spaces, having different curvatures. Each of them is a special case of physical geometry. If one knows in what space–time one develops a theory there is no question about what physical geometry is applicable. But, as discussed earlier, logical geometry is a different type of geometry. It refers to the logical structure of a theory, not to the physical space. "Once a method of measurement is accepted, the question of the structure of space becomes an empirical question, to be settled by observations. ... From the non-Euclidean view of relativity theory, there is no force of gravity in the sense of elastic or electromagnetic forces. Gravitation, as force, vanishes from physics and is replaced by the geometrical structure of a four-dimensional space–time system." (R. Carnap). Note that elastic force is also a convenient theoretical term. We can as well describe the phenomenon of elasticity without reference to the notion of 'elastic force'. "Some physics speculated on the possibility that some day the whole of physics might turn into mathematics. ... In this case, I think they lead to a confusion between geometry in its mathematical sense and geometry in its physical sense. Mathematical geometry is purely logical, whereas physical geometry is an empirical theory." The physical theory of elasticity from our point of view in this book is partly empirical and partly logical as any other well-developed physical theory ought to be.

> The great simplicity of Einstein's equations for moving bodies and light rays is certainly in favor of his claim that non-Euclidean approach is favorable to the Euclidean one, in which it would be necessary to complicate the equations by introducing new correction factors. But it is still far from the discovery of any sort of general principle that will tell how to obtain the greatest overall simplicity in choosing between alternative approaches to physics. *(R. Carnap)*

7.2.6 Kant's Synthetic a Priori

In Kant's terminology knowledge that needs support by experience is *synthetic*. Knowledge that does not require support by experience and is logically self-consistent is analytic knowledge. All analytic statements are a priori true in themselves and independent of experience. A posteriori knowledge is knowledge that requires experience as a reason for asserting it. Kant's question and point of view is that some synthetic empirical knowledge can be as well a priori, not requiring empirical confirmation. "Geometry provided Kant with one of his chief examples of synthetic a priori knowledge." And further "On the other hand, Kant continued, the theorems of geometry tell something about the world. In geometry, Kant was convinced, we have a paradigm of the union of synthetic and a priori knowledge." Carnap continues "From a modern point of view, the situation looks quite different. Kant should not be blamed for his error because, in his day, non-Euclidean geometry had not been discovered." I do not see an error in Kant's position. The discovery of other geometries did not eliminate Euclidean geometry. Those discoveries do not change the essence of Kant's position. Every physical theory has not only a factual part that needs logical support but also a logical part that needs empirical support; thus it is a priori and a posteriori at the same time. Non-observable concepts are inductive logical, that is a priori, in this context. It is a reality of physical theories that synthetic concepts could be a priori and a posteriori at the same time. In the theory of elasticity the description of elastic deformation requires empirical justification, but that is not sufficient for proving this equation correct. The proposition should be part of an a priori logical system. Then justification can be reached by employing logical analytical knowledge of the domain for the physical function that is, as we discussed, a characteristic part of the tangent function. The limiting characteristic of the physical domain is the rate of change of the function; more specifically, it approximately corresponds to $\tan 76^{\circ}$. This universal criterion does not depend on the character of the physical function or the character of its derivative. It is a criterion of possibility of a stable physical existence of the relations.

7.3 Notes on methodology of science

> In the field of the empirical sciences a scientist, whether theorist or experimenter, puts forward statements, or a system of statements, and tests them step by step by observation and experiments.
>
> *(Karl R. Popper, The Logic of Scientific Discovery, 1968)*

This point of view of empiricism is an oversimplification of scientific procedure, and it does not correlate to our experience of building NLTE. Even if we would disregard the part of the process dealing with the creation of the propositional statement and move on to testing, observation and experiments provide insufficient means for effectively justifying a statement. It was shown earlier that not every statement can be tested experimentally. Analytical–logical statements concerning physical relations need mathematical means of justification rather than empirical. Besides observable terms, which can be tested experimentally, every physical law has also theoretical terms of which the magnitudes are obtained as logical inferences from their conjunction with the observable terms and general laws.

A different opinion is: "The history of science shows beyond doubt that the vital factor in the growth of any science is not the Baconian passive observation but the active questioning of nature, which is furthered by the multiplication of hypotheses as hypotheses." (M.R. Cohen). History probably shows this, but it does not mean that the open uncertain character of scientific research should always be the norm. On the example of the non-linear theory of elasticity it was demonstrated that a definitive logical system allows us to build a definitive theoretical system that may have both logical proof and experimental verification. Multiplication of hypotheses could not produce such a result.

"If our reasoning is correct the meaning of our final result follows from our initial assumption." (M.R. Cohen). If our reasoning is correct and assumptions are correct even then a number of conditions must be satisfied to arrive at a meaningful result. "The nature of the subject matter of logic may be better understood when it is seen to be identical with the subject matter of pure mathematics." (M.R. Cohen). Why would one need two identical subjects? Logic has its purposes, its means to reach those purposes, and its rules and restrictions. Pure mathematics is a scientific tool that has its means, its rules and its restrictions. We may consider them and compare them with the rules of logic but they are different. Logic places restrictions on the rules of mathematics in physics.

"Algebraic proofs are in every respect logical proofs and depend no more on any special element of intuition than does logic itself. What is essential in mathematics is that given a set of premises, the conclusion will follow in all cases. There is no strictly logical difference between pure mathematics and deductive reasoning." (M.R. Cohen). Let us see how this works out in application to physics. A mathematical proof can be a logical proof. However a logical proof is not necessarily mathematical in nature. Even if a logical procedure takes the form of a mathematical equation it places its own logical demand on the mathematical procedure and has its own purposes besides the numerical calculation, while pure mathematics does not have such purposes.

"The categoric assertion of either premises or conclusions involves something more than logic. And it is as impossible to derive physical or psychologic truth from pure logic as to build a house with nothing except the rules of architecture." (M.R. Cohen). It is true that logical structure has no particular meaning unless it is completed with a subject matter. However a categorical assertion as to logical necessity concerning physical behavior can be made. That is what was done while constructing NLTE. If a physical phenomenon that is described with all possible characteristics of that phenomenon points to the existence of a limit for those relations then the domain of the function and its limit is in the interval of rapid changes of the tangent function, $1 < \tan\alpha < 4$. This is quite a categorical assertion concerning physical behavior of bodies that is made based on the following analytical (a priori) knowledge: First, that we can describe the change of a function with a mathematical partial derivative. Second, that the derivative has the values of the tangent to the curve representing the function. Third, the knowledge of the tangent values and the character of its changes. It is worth mentioning that the derivative of a function needs not necessarily be a tangent function but can be any function. Nevertheless, the maximum allowable rate of change of the function should not exceed approximately $C_s = \tan 76^o$. This is a universal criterion for the stability of a physical system.

"The form of structure which logic studies is the system of relations which hold between all possible objects that can be ordered into a system. ... In all sciences the consequences of rival hypotheses, such as those concerning the ether, must be deduced irrespective of their material truth, and indeed as a necessary condition before the material truth can be determined. Though two contradictory hypotheses cannot be true in the material of existential sense, both must be assumed to have determinate consequences." (M.R. Cohen). This is an important rule of logic for reaching a determinate conclusive solution of a problem at hand. In this book the consequences of linear theory of elasticity were studied. It became apparent that elastic behavior that is non-linear in nature was described with linear equations. Such a selection of mathematical apparatus fails to describe the elastic behavior of a structure. The non-linear theory of elasticity takes into account the logic of the elastic behavior of a structure and predicts the elastic limit generated by the geometry of a structure.

"The field of every science consists of the relations of certain constants and variables. The constants need not be enduring substances but may be the invariant laws according to which the changes take place. If the actually formulated laws of our physics can be shown to undergo change themselves, it can only be in reference to something else which is constant in relation to them. This justifies the Kantian contention for priori elements in experience, in the sense that every science must assume some invariant categories." (M.R. Cohen). The constant that is taken as the new invariant criterion in the non-linear theory of elasticity is the rate of change of deformation represented with the tangent function. "Judgment is necessary involved in apprehending the meaning of any proposition. Meaning in general is thus a relation between a thing and something else to which it points or refers." (M.R. Cohen). But what distinguishes true from false judgments?

Judgments – true or false – have to be objective categories of logic. In the classical logic and epistemology the nature of a false judgment is better defined than the nature of a true judgment. If a conclusion is false then the proposition cannot be completely true. In the process of understanding and describing a physical phenomenon we deal with two objective worlds – empirical and logical. Each world has its own means to test the truth of our understanding the phenomenon. The empirical world is represented with the observable characteristics of the phenomenon. These characteristics can be tested experimentally, i.e. measured in their own units. Besides these most obvious representatives of the real world incorporated in the description of a phenomenon there is a representative that connects the phenomenon to its material cause. While a cause (force or energy), if established, may be measurable, nevertheless it cannot be included in the description of the phenomenon in which it is absorbed and manifests itself in different forms. In classical mechanics, the connection between external and internal energy is usually represented by some elusive invisible force, such as the elastic force in the theory of elasticity, and the change of initial geometrical parameters. The inference that such an unobservable force exists is made by means of inductive logic. This was discussed already. We think that a force exists because work is performed when the geometry of a body is restored after elastic deformation. Or, work is performed in a permanent change of the geometry of a body in the case of plastic deformation.

In order to assert a proposition it is not enough to have empirical verification. The reason for this is not that empirical data may vary from one experiment to the next, but that a complete description of a phenomenon should include a deductive conclusion from the propositional statement and this logical structure needs to be tested, which cannot be done empirically. Verification of the necessary logical structure can be achieved by establishing its correspondence to a universal geometrical statement describing the relations. In the view propagated in this book on the example of a viable non-linear theory of elasticity such a universal function that provides a possible logical domain for a physical function is a certain interval of the tangent function. The judgment of a proposition in physics can be definitive if it takes into consideration both empirical validation and a system of connected statements that is logically established as true under the umbrella of a universal function/law.

Classical mechanics allows terms to have a double identity. Transformation is made with an equation of equivalence such as Hooke's law, $\sigma = E\varepsilon$. Strain can be represented as stress by using a coefficient – the modulus of elasticity. Both concepts describe the same physical state of an elastic field, i.e. stress corresponding to strain. However, strain (deformation) is an observable entity and can be measured. Stress on the other hand is a theoretical notion. The value of stress can be obtained from Hooke's equation or it can be calculated according to its definition that stress is the elastic force per unit of area of a section.

We have methods to obtain the values of terms but we have no empirical method for identifying and measuring the relation between the physical entities. No matter how obvious and real the relation seems to be, we cannot measure it, feel it with our senses, or observe it with instruments as extensions of our senses. Nevertheless the relations can be identified and proved objectively without involving empirical means. The relations can be tested with consistent patterns of relations that are known to be true. For example, trigonometric functions, such as tangent, cotangent, sine and cosine belong to the objectively known patterns of mathematical behavior that can be used in physics to prove the objective truth of a physical relation between physical characteristics. The empirical and mathematical methods of proof could meet each other at some limit interval of the phenomenon. Divergence of the empirical data from the selected mathematical pattern means that we have no proof for our theoretical structure. In that case we need either to change the pattern in such a way that it describes the phenomenon properly, or to change the physical concepts representing the phenomenon, or to admit failure in understanding the phenomenon.

In our search for discovering true relations in a physical phenomenon we should not rely on observations and experiments as only source of such knowledge. What we 'see' is often our superficial conclusion rather than fact. Such conclusion needs yet logical justification. It can be proven if the relation corresponds to some a priori truth such as a mathematical pattern that corroborates with the facts. By itself an empirical proposition cannot be validated as true. For example, deformation is proportional to the length of a beam. That is a fact. But the same facts can be described in different ways. We cannot tell which description is correct unless the relation satisfies a known mathematical pattern of relations. It is apparent that the so-called empirical law describing deformation depending on the length of a structure and its cross-section is incorrect because it cannot produce a reliable prediction of the change in deformation due to the change of the variables.

John Dewey in "The Theory of Inquiry" says that the principle of the continuum of inquiry gives logical forms to empirical methods. "An adequate set of symbols depends upon valid ideas of the conceptions and the relations that symbolized. Without fulfillment of this condition, formal symbolization will merely perpetuate existing mistakes while strengthening them by seeming to give them scientific standing." (J. Dewey). The idea of formal symbolization is useful for the critique of previous material in the assertion of new ideas that can arise from the critical analysis of existing false beliefs. It is often convenient and useful to present the ideas in symbolic form. This makes it easier for some scientists to analyze false ideas and their connections. All depends on how the scientist used to think. It really does not matter how one comes to the understanding the relations in physical universe. The critique of ideas when clearly expressed in symbolic form allows the advance of new ideas and directions. The common point of view on progress in science is that it is a temporal process evolving as an adjustment to the increasing demands for solving new specific problems. It is also a response to anomalies in the accepted theory. "One of the most important species of anomaly arises when a theory, although not inconsistent with observational results, is nonetheless incapable of explaining or solving these results (which have been solved by a competitor theory)." (Larry Laudan, Progress and Its Problems). An ambiguous character of a theory, lack of proof for its statements, brings imbalance into knowledge. "And yet, on looking into the history of science, one is overwhelmed by evidences that all too often there is no regular procedure, no logical system of discovery, no simple continuous development." (Gerald Holton, Thematic Origins of Scientific Thought, Kepler to Einstein).

7.4 On the nature of a scientific theory

Physical science deals with inquiry into nature. Physical theory is a description of the results of this inquiry. Thus physical theory has a two-fold nature, i.e. physical–empirical and logical–mathematical.

The first step of an investigation is identifying a phenomenon as some cause–effect regularities, then governing the facts related to the phenomenon, classifying the facts and forming the concepts. The next step is describing the quantitative relations between the concepts with the help of mathematical functions. Those functions should satisfy the logical structure of a theory in order to connect that theory to the phenomenon. For establishing the existential value of a theory we have to test the description empirically and logically. In order to communicate the new knowledge a scientist needs to present physical and logical explanations of the theory as a valid scientific description. Logical inductive and deductive methods of justification of a theory are clearly distinct from empirical methods of testing terms and statements of a physical theory, and from mathematical transformations and calculations.

7.5 The theory of elasticity as an organized knowledge

One of the tasks of a theory is to consider the cause–effect relationship. The common observation constitutes the fact that an external force (the cause) applied to a body induces deformation of the body (the effect). This observation has as yet no law attached to it and thus does not have the logical explanation that is necessary for understanding and describing elastic relations. We know two types of valid logical explanation, i.e. deductive and inductive.

Deductive explanation in logic is considered as well defined. Deduction is the process of mathematical inference from the hypothetical statement of proposition to the statement of conclusion. Deduction is based on the assertion that the rules of deduction are correct. Thus, if the proposition is correct, then the conclusion is also correct. If P, then Q. Or, the other form of deductive logic is that if the conclusion is proven false, then the proposition cannot be all completely true.

We cannot, however, establish a connection between cause and effect by means of deductive logic. The cause (external force) is an independent agent. We cannot establish a functional relation between cause and effect for we deal with separate entities. Knowledge of "cause" does not give us knowledge of "effect" yet. A body has its own properties, and the elastic effect depends on the characteristics of the body independent from the external force. The same cause may result in different effects for different bodies. For connecting the cause to the effect we turn to another logic, namely, inductive logic. While in deductive logic we rely on the rules of deduction, in inductive logic we rely on general laws that are accepted as true in the system of knowledge. In the system of classical mechanics there are laws that we are not going to revise for they are at the foundation of the system. Our knowledge of the elastic phenomenon is based on the laws and definitions of classical mechanics. One such general law is the law of conservation of energy. Besides the overall description that an external force deforms a body we may relate the connection of cause and effect to this law of conservation of energy. It will give us a quantitative value in the cause–effect relations. According to this law we may conclude that the work of an external force, which we can determine, is stored as elastic energy in the body. This general law covers under its umbrella a particular relation in the phenomenon of elasticity. Are we certain in this inductive conclusion? We may rely on this conclusion no less than on deductive conclusions. What is true in the selected system of knowledge in general should be true in particular. Note that what is true in particulars is not necessarily true in general.

This definition of the inductive method as a reference to an established general law is not common in logic. Rather, the opposite description is accepted as the definition of inductive logic: if a large number of experiments consistently shows a particular regularity this regularity can be accepted as a general law for such type of relations. As we can see, an inductive conclusion made in such reverse manner does not give certainty and lacks quantitative characteristics. Although some authors consider inductive conclusions as true under statistical laws, no logical inference can be made based on empirical data. Data can be correct and we may use it, though not for the purpose of drawing logical conclusions. An inductive logical inference or transition can be made from a general universal law to a particular phenomenon. Such a conclusion is in the category of logic. If the general law of energy conservation is correct then it is correct for the case of transforming the work of an external force into elastic energy. We do not need confirmation in each particular case.

On the other hand, a statement obtained with the deductive procedure requires logical confirmation. Although the transition from the propositional function to the derivative function is made with the correct mathematical procedure we have no empirical means to confirm that proposition and conclusion are existentially true. The reason is that the derivative function has some pure mathematical properties and therefore cannot be proved or disproved with empirical data. We already discussed this point of view that a pure mathematical statement should have proof of its consistency. We can obtain logical and empirical justification for a physical proposition and its mathematical consequence within a logical system only. The system could be true if it is in the domain of a universal function, and the inference is in compliance with the domain of tangent function in the interval designated for physical stability of the relations. The propositional statement in this system is validated with empirical data. The exclusiveness of the system is linked to the cause–effect connection, or in other words to the balance of work performed by applied forces and the elastic potential of a body. Otherwise the inference that is a part of tangent function is the same for the family of curves corresponding to different forces.

Theory includes descriptions that are, if true, considered as physical laws. Thus, the system of the equation describing the total deformation of a body and the equation describing the rate of change of deformation together constitutes the law of elastic relations.

An analytical description, as was mentioned, contains observable and unobservable characteristics of the deformed body. All characteristics are selected in a way that they relate to the deformation. Deformation itself is an observable and measurable entity. In principle, we have no doubt that in the range of measurement accuracy we have true values of deformation. This truth does not relate to an inductive or deductive conclusion but it is a fact. The deformation of a body depends on its resistance to deformation. The resistance depends on the geometry of the body. The characteristic defined as geometrical stiffness is also an observable entity that can be measured when a special instrument will be available. Until then, it can be measured indirectly by measuring the deformation of a body as response to an experimental force. The main point here is that geometrical stiffness is observable and as such does not require an inductive conclusion as to its true nature.

The external force that causes deformation is also an observable entity. The external force can also be measured, whether it be the weight of another body in contact with the deformed body, the weight of the deformed body itself, pressure of gas or liquid in a chamber, or any other force that is attributed to an independent external entity. We have no reason to doubt their true existential nature. It is not our concern here that the notion 'weight' as the exhibition of gravitational forces is abolished in a system other than classical mechanics. Our concern is that the external force cannot participate in the law describing the elastic relations in a body for the simple reason that it is not part of the body. The force that participates in the description of elastic relations for a body is of different nature. It is the elastic force or stress. The elastic force is an unobservable and non-measurable conceptual entity. We claim its existence because it fits in the frame of our conception and definition of work. In classical mechanics, when work is performed a force should be involved. After removing the external force that causes the deformation, the body returns to its initial position and restores its form. The inductive conclusion is made that there should be some internal forces in a body that, though unobservable, do this reverse job. We can calculate the elastic force because the elastic potential is equal to the work performed by the external force applied to the body. These two inductive conclusions were made on the basis of the general laws for the system.

This logic brought us to the conclusion that the elastic force exists and its value can be calculated. The basic equation of the law then can be written: the deformation is proportional to the elastic force and inversely proportional to the resistance of the structure to deformation. Part of the resistance is a description of the effect of geometrical stiffness on deformation. The coefficient E is obtained on the level of infinitesimal relations between strain (unit deformation) and stress (unit force). This is presumably an adequate analytical description of a deformed unit of material. Here again strain is an observable entity and stress is an unobservable theoretical concept. In order to bring these two concepts into agreement the coefficient E obtained experimentally has been included. Because this coefficient varies for different materials its physical significance attributes to the elastic properties of a material. Our equation is now consistent with the general laws of the system in which it has been designed. It allows us to determine analytically the value of deformation if we know the law of the distribution of unobservable stresses and vice versa.

How we can reach valid conclusions about the existential correctness of relations in the phenomena that we are analyzing? Why do methods of logic give us such possibility? There are two methods of logic, i.e., deductive and inductive. The method of empirical testing for obtaining the magnitudes of terms has no logical meaning in it, nor do we consider here the logic of a measurement. The tests are selected so as to fulfil our logical needs in physics. As well we accept mathematical transformations and calculations without discussing the logic of mathematics. For us, mathematics is only an instrument that we selected according to our logical needs in physics.

The method that is more developed is known as deductive method. With the deductive method, conclusions are based on some selected characteristics of the phenomenon and consistent mathematical rules, providing certainty of results but not their justification. In physics the deductive method is used for obtaining a mathematical propositional statement and inference such as a partial derivative from the proposition. Thus, the conclusion contains the elements that are in the proposition. But the purpose of this procedure is to establish the relation between a function and the selected independent variable concept. Overall, the deductive method provides additional material for analysis and aims at clarifying relations between the variables. Because inference is obtained by rules we do not doubt, classical logic suggests: (a) the conclusion is true if the proposition is true; (b) If the conclusion is not true then the proposition cannot be entirely true; (c) Testing the truth of the proposition is suggested to attainable by empirically testing the conclusion. It was demonstrated and argued in this book that it is impossible to test experimentally the magnitude of a relationship that is the result of inference. The exception in NLTE where an inference can be tested experimentally refers to the geometrical description of a body; this has been discussed already.

The additional requirements for testing the truth-value of a deductive structure and a deductive conclusion are the following. A deductive conclusion in physics cannot be tested and validated empirically if it is separated from a propositional function. The conclusion has the form of a partial derivative that has a ratio as its result. The test of the conclusion can be done mathematically by comparing the relations in the conclusion with a pattern of a geometric or trigonometric function that is known to be true. Such a known correct pattern in the system of plane geometry is the tangent function. The selected mathematical pattern should then be compared with the observable physical behavior. Although observation does not provide proof of the conclusion, the selected trigonometric pattern becomes a reliable instrument for proving the validity of the deductive structure. With this approach we are testing a logical system rather than only a conclusion. The conclusion provides a strict consistent mathematical form for the relations among the components of a statement. These components have physical content. Those physical concepts are in the propositional statement as well. If the conclusion is correct logically then the proposition is also logically justified. Both are in the same domain and obtained with strict mathematical rules that are known to be correct.

Still the question remains whether this description is true to the facts or, maybe, we need to make some adjustments for connecting this system to reality. A good logical–mathematical system should have self-regulatory adjustment. The term to be regulated is an independent variable in an otherwise strict logical–mathematical system. In NLTE the coefficient of elastic stability is such a regulator. The deformation in a real structure can be measured, the modulus of elasticity of the material can be calculated, and the geometrical stiffness of a real structure can be measured. For a similar structure the optimal parameters refer to the rate of change of deformation that we selected, $C_s = \tan\alpha$. The projected optimal geometrical stiffness is a function of measurable components and the coefficient C_s, that is, $R_o = \sqrt{e_a R_a C_s}$. The geometrical stiffness that we obtain with this equation is a theoretical value. By implementing the theoretical R_o in a real structure we eliminate the guess as to how real or accurate the value of deformation is. Its accuracy corresponds to the accuracy of the previous measurements of deformation and the geometrical stiffness of the specimen. Here, for obtaining the value of optimal geometrical stiffness we replace the unobservable theoretical concept of elastic force with the observable and measurable e and R.

As well, the elastic force can be calculated as the mathematical mean value of the forces acting in a body. However, unlike the observable external and internal deformation, the distribution of unobservable stress in a body is a matter of decision with the help of inductive logic. Inductive logic is used for establishing the connection between the cause, i.e. the external force, and the effect that includes elastic force. There is no characteristic common for the body inducing deformation and the deformed body. This means that we cannot describe the relations by deduction. The connection nevertheless exists, and an inductive process may help to establish its character and value. It was mentioned that for induction we use the general laws of the scientific system in which we analyze a phenomenon. These laws are considered as a priori, and we do not doubt them. They should cover the connection of cause and effect. In our case the work of the external force becomes elastic potential energy of a body. We made the inductive conclusion that the total elastic force should have the same value as the sum of the external forces because it performes the same work. However, because the elastic force is a conceptual entity we need a general law that allows us to make an inductive conclusion about its distribution in a structure. Hooke's law of equivalence, namely that in a small unit of a deformed body stress is proportional to deformation, is a general law, empirical and well validated. The distribution of elastic stress should follow Hooke's law. For the current theory of equilibrium of external and internal forces there is no general law. There is a law for the equilibrium of external forces applied to a body that is in equilibrium, but not one for external and internal forces. A quantitative equity exists under the law of energy conservation; we discussed that already. The method of finding stress in a section by equating total stress in a section to forces representing the artificially removed part of a structure has no scientific, i.e. empirical or logical inductive basis. Only in their totality might the internal forces be equated to the sum of external forces.

Also, we deal with observable and unobservable concepts as the elements of a law. The truth-value of a particular observable characteristic is in its measurement. We do not doubt its existential value. The truth-value for the unobservable concepts of a theory is in the certainty of an inductive conclusion. The theoretical acceptance depends on the agreement of unobservable concepts with the general laws of a scientific system. The truth-value of the logical structure of a law depends on the consistency and certainty of the deductive procedure and on the agreement of a system with the universal geometrical function that is serving as domain of physical function. The methodology of construction and verification of a physical theory that is based on definitive logic ensure the certainty of a theory.

The Flow Chart on the facing page presents a schematic representation of the process of designing and justifying the non-linear theory of elasticity.

Flow-Chart Diagram

Logical structure of non-linear theory of elasticity

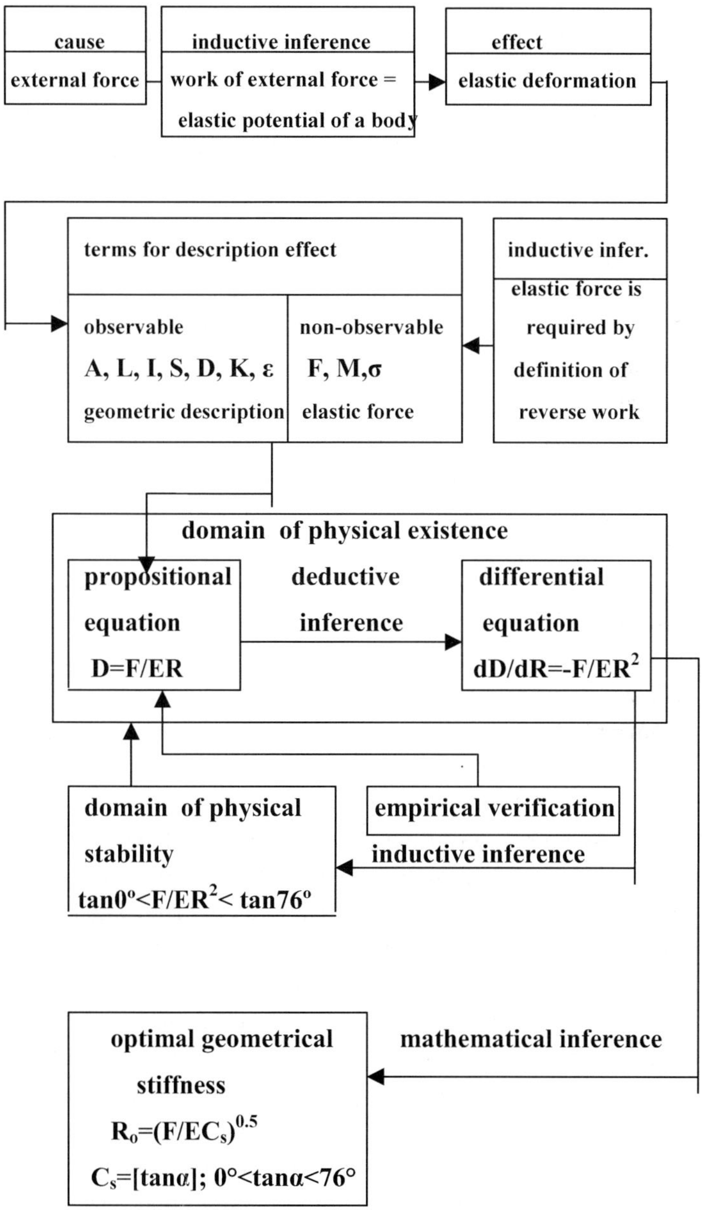

7.6 Logic in mathematics. Commentaries to Bertrand Russell and Kurt Gödel

The influential 20th-century mathematician and philosopher Bertrand Russell in his "Introduction to Mathematical Philosophy" (1918) explained the process of mathematical induction on the example of the series of natural numbers. The question he asked is how the class of natural numbers can be defined. The solution of this question he considers as *mathematical induction*. Starting the description of series with class "0" (empty class) it is possible to describe any number in the series. All properties of class "0" and of the successor of any number belong also to all natural numbers. These properties Russell called "hereditary" or "inductive". Then, the definition that he gives of the series of natural numbers is the following: "We define the 'natural numbers' as those to which proofs by mathematical induction can be applied, i.e. as those that possess all inductive properties." A natural number he defines as the "posterity" of "0" with the respect to the relation of a number to its immediate successor. Russell sees his role in the development of the principle of induction in the importance of defining "ancestor" in term of "parent". In his definition, "The principle of mathematical induction might be stated popularly in some such form as 'what can be inferred from next to next can be inferred from first to last. It is true when the number of intermediate steps between first and last is finite, not otherwise.'" Bertrand Russell comes to the conclusion "It follows that all pure mathematics, in so far as it is deducible from the theory of the natural numbers, is only a prolongation of logic." From his definition of induction it follows that the method of induction can be applied only to finite series. "Mathematical induction affords, more than anything else, the essential characteristic by which the finite is distinguished from the infinite. The principle of mathematical induction might be stated popularly in such some form as 'what can be inferred from next to next can be inferred from first to last.'" And Russell also includes intuition in his understanding of mathematical induction: "When we come to infinite numbers, where arguments from mathematical induction will be no longer valid the properties of such numbers will help to make clear, by contrast, the almost unconscious use that is made of mathematical induction where finite numbers are concerned."

Bertrand Russell discusses the concept of a limit, which is very important for mathematics and physics: "The conception of a limit, which underlies all higher mathematics, is a serial conception."

In physics the concept of limit also refers to series or sets of physical entities. But, contrary to the mathematical limit, which is outside of series or sets, a physical limit belongs to the set.

Concerning the order of a pure mathematical set Russell properly allows consideration of all possible combinations. "In seeking the definition of order the first thing to realize is that no set of terms has just one order to the exclusion of others. A set of terms has all the orders it is capable of." (B. Russell).

In physics though, a set of terms may have different orders but for a given set that is physically true and logically justifiable it can be only one of the orders.

The mathematician Kurt Gödel has changed the logic of mathematics since. In his essay "Russell's Mathematical Logic" Gödel wrote "It is to be regretted that this comprehensive and thorough going presentation of a mathematical logic and derivation of mathematics from it is so greatly lacking in formal precision in foundations . . . ". And further "The matter is especially doubtful for the rule of substitution and replacing defined symbols by their definiens." Of interest to us is Gödel's philosophical view on the possibility of reaching the logical conclusion "As far as the epistemological situation is concerned, it is to be said that by a proof of undecidability a question loses its meaning only if the system of axioms under consideration is interpreted as a hypothetical-deductive system, i.e., if the meaning of the primitive terms are left undetermined."

In 1931 Kurt Gödel published the paper "On Formally Undecidable Propositions" that had great influence on the epistemology and logic of the 20th century. Gödel presents the mathematical proof that in a formal mathematical system there is undecidable proposition such as that neither proposition nor its negation is provable in the system. Such systems Gödel defined as incomplete. "The system S is not complete that is, it contains propositions A for which neither A nor Ā is provable . . . In any of the formal systems there are undecidable arithmetic propositions." Kurt Gödel proved that though it is possible to show consistency of a mathematical system, it is impossible to make a logical judgment about mathematical formulas or systems by means of mathematics within the system itself. This conclusion indeed should not disappoint one looking for a possibility to build logically proven physical systems.

Mathematics is a tool in physics for making mathematical inferences rather than for passing logical judgments. For the latter, different methods of logic are applied to the mathematical structures. It is impossible empirically to prove a physical system. As well it is impossible by means of mathematics to prove its completeness. The judgment on the mathematical systems in physics can be reached with methods that are outside of the mathematics rather than within. The impossibility of such a task has been proven in a strict mathematical manner. However it should be noticed that using logical signs in the presentation of mathematical theorems doesn't provide their logical proof. Logic has its methods, distinct from mathematics.

From all these citations emerges a clear difference between the requirements to the logic of mathematics and the logic in physics. Let us compare the definition of mathematical induction with the definition of physical induction that follows from the analysis of NLTE. Physical induction is an inference from a general accepted law to a particular law. Induction is based on the application of the general laws of the physical system of which the phenomenon is a part. Thus, the inductive inference of the existence of an elastic potential of a body is made by the connection of the general law of conservation of energy to the phenomenon of elasticity. And the induction of the existence of an unobservable elastic force is made from the definition of mechanical work that requires a force. We do not doubt this conclusion in the system that leads to this conclusion. And the conclusion that physical stability exists in a certain interval of the tangent function is an inductive conclusion that is based on the knowledge of the character of the tangent function. As we can see the essence and definition we gave to the physical inductive process are quite different from the definition of mathematical induction given by Bertrand Russell. In physics we cannot make transitional conclusion from one fact to the next unless we know the universal law that covers the facts.

The order plays a significant role both in the logic of physics and in the logic of mathematics. In mathematics the order of magnitude is an important mathematical property. In physics we should correlate the order of magnitude of an independent variable with the order of magnitude of the function. Correlation is important. It is a restricted order.

In geometry the order of points on the line, or lines through a point in a plane, or planes through a line, is important. Russell suggests "In seeking a definition of order, the first thing to realize is that no set of terms has just *one* order to the exclusion of others. A set of terms has all the orders of which it is capable." Thus, natural numbers can be arranged not only in order of magnitude; other arrangements can be made, such as the set of odd or the set of even numbers. The same is true for the arrangement of points on a line. "The resulting order will be one which the points of the line certainly have; the only thing that is arbitrary about various orders of a set of terms is our attention, for the terms themselves have always all the orders of which they are capable." (B. Russell). Thus, while turning to geometry Russell sees the necessary restrictions on order in a specific set: "The order lies, not in the *class* of terms, but in a relation among the members of the class, in respect of which some appear as earlier and some as later." (B. Russell).

Physical relations presented as geometrical patterns of relations have an objective certainty to it. In physics all these seemingly abstract properties of numbers that in their totality describe physical relations represent only one of the possible abstract sets. However, after selecting this set it is proven that the set has objective and existential properties.

The graphical presentation of a function vs. an independent variable gives an order that can be interpreted as the physical behavior of a set, of which the members are physical objects or characteristics. Russell assigns significance to any "two terms in the class which is to be ordered, that one 'precedes' and the other 'follows'." Russell in his book distinguished three properties the ordering relation should have: (1) A relation in a set has direction, x_1 precedes x_2. This property of order is called *asymmetrical.* (2) If x_1 precedes x_2 it must precede x_n; this relation is *transitive.* (3) Of two points on a line, one must be to the left of the other; this property is called *connected*. From the point of view of mathematics according to Russell "When a relation possesses these three properties, it is of the sort to give rise to an order among the terms between which it holds; and whenever an order exists, some relation having these properties can be found generating it."

It was previously discussed that in physics for establishing the existential order in a set it is not enough to have the magnitude of the members. There should be a logical procedure for selecting the members of a set. The three rules for a mathematical set are not sufficient for establishing physical order. Thus, deformation of a bar can be presented as a function of its diameter, or as a function of its area of cross-section, or as a function of its geometrical stiffness. All three cases are in compliance with the mathematical three-rule order, but only one of these functions has existential order.

According to Russell, the mathematical relation of practical importance is the generation of series by means of a three-term relation. In geometry it is that one of three points on a straight line lies between the two others. The notion "between" is a characteristic of open series real order as opposed to "cyclic" series.

Russell asks the important question of how to *construct* relations having some useful property by means of operations upon relations which only have rudiments of the property. The answer he sees in the magnitudes that have relation of greater and less. This is a fundamental asymmetry as a property of relations.

For physical relations the magnitudes of the terms are not sufficient for constructing meaningful useful relations. For example, compare the deformation of two bars with different cross-section diameters. The magnitudes of deformation by themselves are not sufficient for establishing the domain for the set of such structures.

It is impossible by mathematical means only to prove a physical theory. There are general laws of physics and the inductive and deductive methods of definitive logic that allow justify a physical theory.

7.7 On explanation of a physical theory

The philosophers of science classified the scientific explanation of a phenomenon with respect to the logical structure of a theory. One of the problems is that a physicist invents a physical theory while a philosopher gives a logical explanation. Neither does the philosopher construct a theory according to his logical rules, nor does the physicist use those rules for constructing his theory. When I tried to explain NLTE it became apparent that different rules of logic need to be introduced for successfully explanating and ultimately for constructing a physical theory. Explanation in physics is not only the explanation of a logical matrix. The physical content of a theory must be explained as well. For example, the logical explanation that the limit of a set of structures described with the equation of elastic deformation is in the interval $1 < dy/dx < 4$ (tan 76°) because the rate of change of the function beyond tan 76° increases drastically is still incomplete. Facts suggest that the limit of elasticity may have different origins, and may depend on the material of the structure as well. Thus a physical explanation is offered. The limit of elasticity of the material differs from the limit generated by the geometry of the structure. The lesser of these is the actual limit for a structure. This conclusion, although it has the logical step of comparing values of limits, is impossible to reach without the empirical physical finding of the existence of limits of different origins. Thus, explanation in physics has a complex character. The general principle of uniformity suggests that explanation in any science should be of complex nature. The material on explanation in science presented here follows the book "Scientific Knowledge" by Janet A. Kourany.

7.7.1 Inferential conception of explanation

Inferential conception of explanation follows the classical logic argument. If one believes that the set of statements called premises are validated and believed to be true, then one should believe that the conclusion that follows these statements should also be true. "Giving the logical relation of *validity* holding between the premises and the conclusion [of an argument], and assuming that the premises are true, we have strong, but not conclusive, grounds for believing the conclusion." (Janet A. Kourany, "Scientific Knowledge"). In this classical approach one thing at least remains logically unexplained: what are the criteria for premises to be considered as "true"? There is no sufficient logical ground given for believing in premises and assertion conclusion.

"According to the inferential conception of scientific explanation, scientific explanations are deductive or inductive arguments – logical inferences. These arguments are composed of (1) a conclusion, called the *explanandum*, describing the phenomenon (event, fact, law) to be explained and (2) a set of premises, called the *explanans*, describing the facts and laws provided to account for explanandum. … Scientific explanations have been divided into three classes: 1. Deductive–nomological (D–N) that is, deductive explanations that include only universal laws in their explanans. (2) Deductive–statistical (D–S) explanations; that is, deductive explanations that include at least one statistical law in their explanans. (3) Inductive–statistical (I–S) explanations; that is, inductive explanations that include at least one statistical law in their explanans." (J.A. Kourany).

The American philosopher of science Carl Gustav Hempel proposed a model of scientific explanation that refers to an argument that has as its premises a set of statements, the explanans that describe the initial conditions and contain at least one scientific law. The argument also has at least one sentence, the explanandum, that deductively follows as its conclusion. Hempel gives a set of requirements for scientific explanation. The explanandum must be a deductively logical consequence of the explanans. The explanans must contain at least one general law necessary for the derivation of explanandum. The explanans must have empirical content; i.e. it must be empirically testable.

Hampel believes in the sufficiency of the empirical conditions of adequacy. These conditions are as following. The sentences constituting the explanans must be true or the explanans have to be highly confirmed by all the relevant evidence available rather than that they should be true. Probably, thist means that explanans statements should be supported empirically. We discussed already that support by facts is a necessary but not sufficient condition for assertion that a proposition is 'true'.

Hempel's inductive–statistical model of scientific explanation is parallel with deductive–nomological explanation. The difference between them is that the I–S requires use of a statistical law from which the explanandum is inductively derived from the explanans. Here the assumption is made that all readers know and agree with some definition of inductive inference. In fact we can find in the literature different definitions of inductive inference. The most common understanding is that generalization from instances to law is an inductive leap. Such generalizations are only probable. Laws are central to Hempel's models of scientific explanation. He did not give a satisfactory definition of what constitutes a scientific law. Hempel makes a distinction between fundamental and derivative laws. According to Hempel's D–N model, a good explanation is one that makes the occurrence of the event a certainty. The difference between prediction and explanation is, in his opinion, that the argument relative to the event-to-be explained is prediction when preceding it, and explanation when following it. In fact, prediction and explanation of an event are different in essence. For example, the prediction of the limit of elasticity in the interval of tangent to the curve describing deformation $1 < \tan \alpha < 4$ is mathematically obtained. The explanation that the limits are relative values occurred after experiment showed that some times the mathematical limit is true but other times the actual limit is less than mathematically predicted. Because the formula for deformation does not include the limit of the material, it is this limit of the material that sometimes interferes with our mathematical prediction. As we see the prediction and explanation can be two different types of approaches. Wesley Salmon proposes one alternative view on explanation. Salmon's point of view accounts for the role of causality in most scientific explanations.

Nancy Cartwright in "The Truth Doesn't Explain Much" pointed out a class of phenomena that has no law to describe it and nevertheless is true under special circumstances. Most models of explanation offered recently in the philosophy of science are covering law models. Any law does not cover many phenomena, which have perfectly good scientific explanations. "They are best covered by *ceteris paribus* generalizations – generalizations that hold only under special conditions, usually ideal conditions." (N. Cartwright). In her opinion "Most scientific explanations use ceteris paribus laws. We are lucky that we can organize phenomena at all. There is no reason to think that the principles that best organize will be true, nor that the principles that are true will organize much." (N. Cartwright). She does not explain what she means by "true". Science is impossible without organization of description of phenomena. The explanations in science are applied to the series of events rather than to the singular occasion. Ceteris paribus explanations, as described by Cartwright, can be referring to the class of non-scientific explanations.

Scientific knowledge would not advance far with such attitude. But this does not mean that we should not offer plausible explanations for phenomena we cannot describe by a law, or to offer explanations for restricted laws. In my opinion we create the laws in a manner that fits our purposes and can be explained.

In "The Pragmatics of Explanation" Bas C. van Fraassen writes, "There are two problems about scientific explanation. The first is to describe it: when is something explained? The second is to show why explanation is a virtue. Presumably we have no explanation unless we have a good theory."

The explanation of the assumption that the equation $e=N/ER$ is valid has an inductive–deductive basis. It has empirical validation. The equation is mathematically constructed on the basis of the classic numerously tested equation $e=NL/EA$. This is done by substituting the ratio of the geometrical parameters (A, L) by the characteristic of geometrical stiffness R, $KA/L=R$. The experimental coefficient of specifics K can be obtained with the help of statistical law. Still the equation of elastic deformation is an inductive conclusion because the unique physical correctness of this statement cannot be validated with empirical means only. The logical structure that can be verified or falsified should include mathematical inference from this empirical statement presented in mathematical form for examinating the dynamics of the relations between the components. It is important for making an adequate physical conclusion. For validation of the empirical statement we have to validate the system of the basic and derivative equations. The derivative equation, though obtained with a deductive procedure, also needs inductive validation. An inductive conclusion connects both statements with the tangent function in the intervals of rapid and slow changes. The logical structure finds its validation in the character of the universal tangent function and an experiment that was designed to test the conclusion. The classic equation $e=NL/EA$ was changed because, as discussed above, there was no correspondence between the mathematical description and the physical phenomenon in its domain. Empirical validation of the system of basic and derivative equations is necessary, but it is still not a complete validation of the system. When the external force is constant the function has the character of a hyperbola.

The derivative in this case can be identified with the tangent function. Thus we not only assign concrete certain values for the results of rate but for the results of deformation corresponding to the mathematically known rate as well. Also, the same degree of empirical validation that is obtained by testing the basic equation is applied to the derivative equation. When the external force is dynamic, for example cyclic, the deformation changes correspondingly in time. However the criterion for the change of deformation is still connected to the tangent function and still should not exceed $\tan 76^{\circ} = 4$.

The validation of a system and the explanation that follows is achieved with the inductive–deductive correspondence of the interconnected statements. In the current classification, the inferential conception of scientific explanation is a deductive or inductive argument.

7.7.2 The causal conception of scientific explanation

The causal explanation is referring not so much to the logic of a theory but rather to the explanation of the internal physical mechanism of phenomenon. "Explaining the world and what is going on in it means, accordingly, laying bare its inner working, its underlying causal mechanisms." (J.A. Kourany).

A distinction should be made between the explanation of the world and the explanation of our scientific theories and their connections to a phenomenon it describes. The latter, rather than the explanation of the world, is the purpose of science.

The causal explanation is based on assumption that by finding and explaining the cause of a phenomenon we explain the phenomenon. In his essay "Why Ask, 'Why?'" Wesley C. Salmon expresses his point of view on scientific explanation. "Scientific knowledge is descriptive – it tells us *what* and *how*. If we seek explanations – if we want to know why – we must go outside of science, perhaps to metaphysics or theology." And further "It is now fashionable to say that science aims not merely to describing the world – it also provides *understanding, comprehension*, and *enlightenment*. Science presumably accomplishes such high-sounding goals by supplying scientific explanations. The current attitude leaves us with a deep and perplexing question, namely, if explanation does involve something over and above mere description, just what sort of thing is it?" Explanation is certainly more than description. Explanation involves the logic of connections among terms and statements. And explanation should involve the physical content of a theory.

Salmon then criticizes two main directions in scientific explanations. His opinion in abbreviated form is as follows: "1. The first of these intuitions is the notion that the explanation of a phenomenon essentially involves *locating and identifying its cause or causes*. I shall call the general view of scientific explanation, which comes more or less directly from this intuition, the causal conception. 2. The second of these basic intuitions in the notion that all-scientific explanation involves *subsumption under laws*. According to this view, a fact is subsumed under one or more general laws if the assertion of its occurrence follows, either deductively or inductively, from statements of the laws. Since this view takes explanations to be arguments, I shall call it *the inferential conception*." (W.C. Salmon).

Both of these conceptions of explanations have their successful place in scientific explanation. Thus, in the theory of elasticity the magnitude of the internal elastic force can be found under a "covering law", that is, the universal law of *energy conservation*. The sum of internal forces is equal to the sum of external forces. The elastic potential energy stored in a body is equal to the work of external forces performed on the body. Inferential conception, in my opinion, is an argument supposed to give an explanation that is necessarily true, but in fact it needs to meet some additional conditions. This was discussed already in the preceding section concerning inferential conception on the example of law of elastic deformation. The scientist can tell that the mathematical inference that describes changes of the essentially empirical law in the statement of premises is true if the total argument gives the adequate description of the physical phenomenon. In physics that description can be considered deterministic. What this means is that if premises and the inference–conclusion are empirically validated and if the logical structure of the conclusion is proven by identifying it with a universal logical statement, then this argument is an explanation of a description of a physical phenomenon. I called such explanation tentatively inductive–deductive for it has partly an inductive explanation as to correspondence of the statements to a universal law and partly a deductive explanation, that is, the explanation in the strict frame of logical–mathematical structure. Salmon's point of view is "If, as Laplace believed, determinism is true, then every *future* event would thus be amenable to deductive–nomological explanation."

Determinism is a true feature of scientific law. However the explanation of a scientific law is inductive–deductive. If a scientific law describes an inductive–deductive connection from the facts to the conclusion then it is a deterministic law. The deductive part is deterministic for it follows deterministic rules. The inductive part is deterministic for it follows general law. "Causal processes and causal interactions are, of course, governed by various laws, e.g. the law of conservation of energy and momentum." (W.C. Salmon). It seems that Salmon considers inductive inference from a general law to a specific law as deduction. "Near the beginning, I suggested that deduction of a restricted law from a more general law constitutes a paradigm of a certain type of explanation. No theory of a scientific explanation can hope to be successful unless it can handle cases of this sort." (W.C. Salmon). And further – "The deductive relations *exhibit* what amounts to a part–whole relationship, but it is, in my opinion, the physical relationship between the more comprehensive physical regularity which has explanatory significance." In the present work, explanation from general law to specific law is understood as induction rather than deduction because it has no rules of transformation characteristic for deduction.

The part–whole relationship is important to the overall explanation of scientific theory. The concepts of scientific theory, argument and the parts of argument and whole–part relationship – all are units of construction of scientific theory and tools for the explanation of a theory as well.

7.7.3 The erotetic conception of scientific explanation

> Any request for a scientific explanation can be analyzed as a why-question, and any scientific explanation can be analyzed as an answer to a why-question. Since the logic of questions is known as erotetic logic, this approach can be called the erotetic conception of scientific explanation. *J.A. Kourany*

Does this approach help to explain scientific laws, or may it, at least in some cases, interfere with the explanation? Let us state the question:

Why is the physical characteristic of a whole considered to be identical to the corresponding characteristic of its infinitesimal part? Stated in the form of a why-question this raises more questions than answers, and hardly helps to clarify the basic relations between a whole and its parts. First a question to this why-question: Is it a common belief or scientific knowledge? This is not a why-question. But suppose that we have facts enough to believe that in some instances in the properties of a whole investment is made by the parts. What kind of explanation to that fact is required? The erotetic explanation methodology does not add significantly to the methodology of science.

7.8 Theory and observation

How a theory is connected to observations is an important practical and philosophical question. There are different definitions of scientific theory. "Structurally, theories are sets of statements some of which state laws, others of which are singular factual or existential claims. Furthermore, theories contain some terms that refer to unobservable entities or properties. Theories explain not a particular law or phenomenon but whole ranges of each – this range typically called the domain of the theory." (J.A. Kourany).

The point of view in this book is that the connection between theory and observation is achieved both through logical–mathematical structure and through observable and non-observable terms of a theory. The observable terms by definition are connected to observation and measurement. Unobservable terms that also have physical meaning are connected by means of inductive inference through general laws to the a priori observations. The correspondence between the pattern of mathematical description of a function and the character of the physical relations needs to be tested. Thus the description of elastic relations in NLTE is in conformity with the experimental findings in the intervals of slow and rapid changes of deformation.

Thus, the statement $e = N/ER$ contains both observable and unobservable concepts. Deformation e and geometrical stiffness R are observable entities – concepts that can be measured. On the other hand, internal elastic force is an unobservable theoretical concept. We do not see the elastic force and we cannot measure an elastic force or stress. Nevertheless this unobservable force is an objective entity. It is an inductive inference from the fact that after we remove an external force applied to the body the elastic force restores the geometry of the body. Also, existence of an elastic force is covered by the general law of energy conservation. The work of an external force that is spent on deformation is stored as elastic energy of the body. From the point of view of Hooke's law, internal elastic force is proportional to deformation. Although the internal elastic force is an unobservable entity, the external force is an observable and measurable entity. Both the observable external force and the unobservable forces of the elastic field are real in the system of classical mechanics. In the derivative equation of elastic deformation, $\mathrm{d}e/\mathrm{d}R = -N/ER^2$, we also deal with observable and unobservable terms. The rate of change of a function is a theoretical term. We have no conventional means to measure it. But it is a necessary objective characteristic of a function describing physical relations. According to this research the rate determines the domain of a function and the limit of physical existence. The law that covers the existence of the physical phenomenon is the tangent function in the interval $0 < \tan \alpha < 4$. In NLTE an experiment shows change of deformation at some constant particular rate that increases in the interval of elastic failure and thus may serve as characteristic of the boundary of elastic behavior. Deductive inference gives us the equation describing the rate.

Stephen Toulmin in "The Ideal of Natural Order" connects explanation to the cause of a phenomenon. "By bringing to light causes of the appropriate kind, e.g. Newtonian 'forces', we may reconcile the phenomenon to the theory; and if this can be done we have shall achieved our "explanation". Every step of the scientific procedure that requires explanation is governed by logic of inductive and deductive explanations. "Scientific discoveries, however, do not consist in arguments, which are plausible *ad hominem,* but rather in explanations, which will stand on their own feet. In these explanations, the relation between the 'familiar' and the 'unfamiliar' may be reversed." (S. Toulmin).

Scientific theory within an appropriate scientific system should be true to the system and logically consistent within its own structure. The explanation is needed to show the consistency of a theory. What should be explained?

(1) Connection of a theory to the general universal laws in the system. Thus, NLTE obeys the laws of classical mechanics such as the law of energy conservation. The work of external forces acting on a body is stored as elastic energy of the body. NLTE does not question this law. The law is *a priori* for the theory.

(2) NLTE is in an agreement with the logical structure of Euclidean geometry. Specifically, NLTE uses the tangent function for representing inference from the propositional equation. The tangent function is accepted as a known function of certain independent relational values that can be accepted as *a priori*: $y/x = \tan \alpha$ is consistent within Euclidean geometry and not questioned in the NLTE. The properties of the tangent function have been used for describing changes that occur in the elastic relations and ought to be described with a propositional physical equation. The tangent function can be presented in two intervals: the interval of slow changes, $0 < \tan \alpha < 1$, and the interval of rapid changes, $1 < \tan \alpha < \infty$. Any smooth continuous function that occupies the domain of the tangent function has changes that are described logically with the tangent function.

(3) When we deal with some cyclic dynamic forces and deformation cannot be described with a smooth function, the limiting characteristic for the rate of change of deformation is nevertheless the same value, that is, approximately $\tan (75–76)^{\circ}$. It is a universal coefficient of elastic stability that sets the limit for the physically stable state of existence of any particular system before changes of its internal structure occur.

(4) The physical phenomenon in question is the object of cause–effect explanation. For the theory of elasticity there is a basic simplest explanation. An external force (the cause) applied to a solid body under a certain condition, i.e. that the body has supports preventing its movement, causes a change of geometry of the body (effect). The explanation of cause-effect connection is inductive. The change of the body is accompanied with exchange of energy. External forces perform a work of deformation that is stored as an elastic energy

(5) The explanation of why a body can be deformed to some extent without destruction refers to *the effect* only. All bodies of all materials have elastic properties. There is no logical explanation for this effect. We accept it as it is. The explanation can be only a speculation, that parts constituting a body have internal forces, depending on the geometry of the internal structure, that keep these parts together. The forces keep parts together until the energy of the external forces exceeds these internal forces and permanent changes or destruction of a body occurs.

(6) The internal forces are unobservable entities. This does not mean that the elastic force is a non-objective term, but that it is in principle impossible to see and measure it directly or indirectly. The only way to obtain the magnitude of the elastic force is by theoretical means. It is an inductive conclusion in our system of knowledge that such forces must exist. The inductive argument leading to this conclusion is the following: After removing the external forces the body returns to its initial form, size and position, which means that work has been performed. The explanation under the law of conservation of energy is that the work of the external forces was stored as elastic energy. The definition of work in our system of knowledge is that if work is performed then there must be a force. This means that we have an elastic unobservable force that performs the work of returning a body to its initial shape/position. Although the law of elastic change can be written without a term for the elastic force, the description that includes the elastic force is a convenient objective description.

(7) We should explain the domain of the law of elasticity. It appears from our research that a body or a system of bodies has the ability to change its geometry to some limited extent. The limit of elastic relations appears to be the constant criterion of elastic stability, approximately equal to $\tan 76^{\circ} = 4$. A physical system may exist only at certain finite conditions. The character of the tangent function suggests this value as a physical limit. Above, the tangent function rapidly goes to infinity.

(8) We have to explain the selection of the entities we use to describe the phenomenon. Observable entities can be measured. Unobservable entities should be explained with an inductive procedure of

compliance with a universal law. All observable and unobservable entities should satisfy the logical structure of a theory.

(9) We have to explain the logical structure of the theory. This structure in NLTE includes the basic equation of deformation and the derivative equation for the rate of change of deformation. Both equations should occupy the domain of the tangent function in the interval $0 < \tan\alpha < 4$ for possible physical existence. Beyond this interval is the condition of infinity that is incompatible with material existence.

(10) We explained why the function should be non-linear and should occupy both intervals of the tangent function. Elastic relations have a limit. The only way to describe such relations and to determine the limit is with a non-linear function in the interval of possible stable physical existence.

7.9 Validation of scientific theory

Up to the present time we still speak not about proof or confirmation of a physical theory but of its validation; this carries a much lesser degree of assurance in our scientific endeavor. It can be explained with our understanding of the term 'proof'. In our empirical philosophical attitude to the external world, proof is identified with finding a fact that is identical in essence to the theoretical term or concept. But a theory cannot be proved with particular facts, because some non-observable parts of a theory cannot be connected directly to the facts. Thus proof involves the assumption that the connection of observable facts is true evidence of the existence of certain unobservables. And the other part of a theory, i.e. the connection among terms, is in principle non-factual but logical. Both parts, factual and logical, can be presented in a mathematical system of functions. This system could be proved both logically and empirically. We use such methodology for proving the consistency of the non-linear theory of elasticity. In this methodology logic is somewhat different from classical logic. Although many philosophers of science have been identifying logic and mathematics, mathematics has consistent theorems that can be proved. Logic, on the other hand, has no similar procedure that would prove logical argument. This is what needs to be introduced in logic. In view of an empiricist – "It is commonplace that scientific theories are tested by the results of observation and experiment. Indeed, that is the hallmark of the empirical method, which is held to be a primary source of the success of science. Hence scientific theories are tested by deducing from them consequences regarding observable state of affairs and then comparing these consequences with the result of observation." (J.A. Kourany). Validation of a theory according to empiricism can be reached only through observation of the phenomenon, direct or indirect, and experiment. However, because we cannot observe the internal mechanism of a phenomenon we first make an assumption. From this assumption we draw a conclusion about possible results. Then we design an experiment that would support such outcome. On the way of such connection of a theory with the facts many doubts remain. Different mechanisms may lead to the same end result. We still do not know and cannot test it. The experiment, as a rule, is designed in order to obtain results. The reasons for such approach are not only psychological but practical as well. Also, even if we have several assumptions at a time, the experiments should be different for different assumptions. If we fail to design an appropriate experiment then not only our assumption fails but our methodology fails as well. This is despite Popper's suggestion that it is necessary to design an experiment that deems a hypothesis for failure. It is impossible to deny the importance of observation and experiment. But one should not expect that observation would lead to a theory. And the value of experiment depends on the value of a theory. We have to build a theory using a consistent logical scheme proven as necessary. The empirical validation of such theory is necessary. It can bring to the theory new facts and ideas.

What is the relation to the theory of elasticity of this point of view on the role of scientific theory in explaining observation and connecting theory and observation? Actually, in the case of the non-linear theory of elasticity my expectation was that the partial derivative of the standard equation of deformation should point out an increase in deformation rate in the interval of rapid changes of the tangent for the derivative from the standard equation. The expectation has not proved to be correct and furthermore appears to be impossible with the classical description of deformation. This was the reason to change the formula so as to fit the logical expectation. The new explanation led to new facts. The limit of elasticity may be of different origins. Experiment showed that the limit can be predicted from the new equation of deformation, or it can be the experimental limit of the material, depending on their relative values. The theory allows us to predict the limit. It had first deductive validation and then empirical validation. The observation, inductive and deductive conclusions by themselves cannot explain and validate a theory. Both methods, inductive and deductive, are involved in constructing and validating a theory. Every scientific statement has both observable and unobservable concepts. The term "unobservable" does not simply mean that we have no technical means to observe a characteristic. It rather means that a concept that in a theory conveniently can describe a phenomenon is not present as an entity that can be measured directly or indirectly without the help of the inductive method. Thus, in a body deformed by external forces the internal force/stress cannot be measured. However, we have reason to make the inductive conclusion that such force should exist according to the system of classical mechanics. The fact that, after removing the external force acting on the elastically deformed body, the body returns to its initial form and position makes us conclude that there is an internal force that does this job. The general law of energy conservation allows us to calculate this internal, principally unobservable force. Is this indirect measurement? No, it is a theoretical procedure that is based on the inductive inference from our general conception of how things occur in nature. The force can be an observable concept and in such case it can be measured. Thus, external forces that cause deformation are observable in this phenomenon (can we see them or not) and we can measure the external forces directly or indirectly. I think that the definition of unobservable force can be the following: an unobservable force is a force that is considered a characteristic of some field (e.g., gravitational, elastic, electromagnetic).

What kind of reasoning has the philosophy of science used for the support of a theory?

> Regarding such questions there has been – and still is – much disagreement in the philosophy of science. In fact there are no fewer than six major approaches to such questions of theory testing: (1) justificationism; (2) falsificationism;

(3) conventionalism; (4) the methodology of scientific research programmes; (5) Thomas Kuhn's sociological approach; and (6) the testing paradigm of scientific inference. *(J.A. Kourany)*

7.9.1 Justificationism

The point of view common in justificationism has been discussed at length in this section and in our comments on work by Rudolf Carnap (section 3.2). In this approach a theory cannot be verified because there are many observations of consequences that can correspond to the unobservable entities and to the relations because it involves a hypothesis that cannot be verified. "According to justificationism theories can never be "verified" – that is definitely shown to be true. They can, however, still be partially justified or 'confirmed' to different degree." (J.A. Kourany).

7.9.2 Falsificationism

The philosophers who hold this point of view believe that the confirmation of a scientific theory is impossible. The observation of facts connected to a phenomenon does not yet give us the right to assert that the theory correctly represents the phenomenon. On the other hand, if the theory including its consequences is not supported by facts this proves the falsehood of the theory. Theories that are not eliminated in such vigorous testing process can be accepted temporarily until a fact contradicting the theory is found.

We discussed already that although the observation of facts by itself does not constitute confirmation of a theory, in conjunction with a proven consistent logical system for the theory they justify the theory in an exclusive manner, or, we may say, prove the theory.

7.9.3 Conventionalism

Conventionalist thinks that the scientist should not rely on empirical grounds for a theory because facts can be adjusted to any conceptual framework. The theories usually are chosen by convention on the basis of number of considerations.

7.9.4 The methodology of scientific research programmes

This has a similar position as conventionalism, with some adjustments. Here any theory can be saved from negative experimental results by suitable additional hypotheses. This is the case for many scientific theories.

7.9.5 The testing paradigm of scientific inference

According to this methodology scientific theory can be represented by a theoretical hypothesis. This hypothesis should take the form of a model that includes elements of the real system of objects. If such scientific model in a process of physical testing demonstrates high probability of a positive outcome then the hypothesis is true. If, on the other hand, the model demonstrates high probability of negative outcome then the hypothesis is false. "According to Giere, the unit of appraisal in science is a theoretical hypothesis – that is, a construct that identifies elements of a theoretical model or defined type of system with elements of a real system of objects and then goes on to claim that such a real system of objects exhibits the structure of the model." (J.A. Kourany).

7.9.6 Summary

All these approaches share certain features in common. They all agree that tests need to be conducted for the observable consequences of a proposed theory. The compliance of the theoretical predictions with the results of testing may justify the theory for the time being. None of approaches gives criteria for distinguishing scientific approach, true or false, from non-scientific approaches. Although none of these approaches for validating a scientific theory can be accepted as a whole and exclusive method of validation, each of them has some features that can be used in validation. There is objective value in the inductive method and in the deductive method of validation. It is good to be critical towards a scientific theory. It is important to have a model of the theory and test this model.

Let us take a short excursion to the writings of main proponents of different methods. The work of Rudolf Carnap, who is a strong advocate of empirical confirmation, has been discussed at length in this book. Karl Popper is a major proponent of falsificationism. He stated questions clearly. But he did not answer these questions satisfactorily. His questions are following. "When should a theory be ranked as scientific?" or "Is there a criterion for scientific character or status of a theory?" Popper's answer to these questions is "One can sum up ... by saying that the criterion of the scientific status of a theory is its falsifiability, or refutability, or testability." This paradox may contain a gram of truth, for metaphysical or religious theory cannot be falsified for it has no logical structure in its foundation that can be proved or disproved. But in order to be a really useful, the methodology of validation should concern with a practical schema for *validation*, not just for *falsification* of a theory. The methodology should have both a critical part and a positive part, and a descriptive procedure for exercising both while applied to the same material. Popper explains his criterion of falsifiability as a criterion for distinguishing scientific from pseudoscientific theory. "The problem which I tried to solve by proposing the criterion of falsifiability was neither a problem of meaningfulness or significance, nor a problem of truth or acceptability. It was the problem of drawing a line between the statements, or system of statements, of the empirical sciences, and all other statements – whether they are of religious or of metaphysical character, or simply pseudoscientific." (K.R. Popper). This criterion of demarcation Popper identifies with the criterion of testability, or falsifiability, or refutability. What does it mean? "The critical attitude, the tradition of free discussion of theories with the aim of discovering their weak spots so that they may be improved upon, is the attitude of reasonableness, of rationality." This is a good pastime and can be useful, but discussion of a theory cannot substitute for methodological criticism of scientific theory. On the other hand, Popper believed that logical proof for a theory is impossible. He writes: "The demand for rational proofs in science indicates a failure to keep distinct the broad realm of rationality and narrow realm of rational certainty: It is an untenable, an unreasonable demand."

On the example of non-linear theory of elasticity it was shown that we do have means to reach rational certainty. Karl Popper, who is always very assertive in his own propositions and statements in contrast to his teaching of scientists, states: "The critical attitude may be described as the conscious attempt to make our theories, our conjectures, suffer in our stead in the struggle for the survival of the fittest. It gives us a chance to survive the elimination of an inadequate hypothesis – when a more dogmatic attitude would eliminate it by eliminating us." Here is a very colorful analogy to the evolutionary process of survival of the fittest. Science adapted different approaches to its progress so that it should not have the pace of a natural evolutionary process that retains its errors and unnecessary information for millions of years. Popper identifies the process of validating a scientific theory with induction in the common understanding of the term induction. But his fixation on falsification of a theory rather than on validation returns him to the problem of demarcation. "It is obvious that there [is] a close link between the two problems: demarcation and induction or scientific method. It was easy to see that the method of science is criticism, i.e., attempted falsifications." (K.R. Popper).

Criticism is present in science but it is not a main constructive approach to problems. "Why, I asked, do so many scientist believe in induction? I found they did so because they believed natural science to be characterized by the inductive method – by a method starting from, and relying upon, long sequences of observations and experiments." (K.R. Popper). There is no consensus on the definition of inductive method. The inductive method as defined in this work applied to the problems in scientific theories connects specific entities and relations with general principles or laws. The inductive method connects the related non-observable entities to the observable entities empirically in order to find the magnitude of non-observables that cannot be tested empirically otherwise and thus finding empirical support for unobservable entities under application of universal laws. "Of course, one can invent new problems of induction, different from the one I have formulated and solved." (K.R. Popper).

Popper summarizes his conclusions as follows:

"1. Induction, i.e., inference based on many observations, is a myth.

2. The actual procedure of science is to operate with conjectures: to jump to conclusions – often after one single observation.

3. Repeated observations and experiments function in science as *tests* of our conjectures or hypotheses, i.e., as attempted refutations.
4. The mistaken belief in induction is fortified by the need for a criterion of demarcation which, it is traditionally but wrongly believed, only the inductive method can provide.
5. The conception of such an inductive method, like the criterion of verifiability, implies a faulty demarcation.
6. None of this is altered in the least if we say that induction makes theories only probable rather than certain."

Popper's statement that "Only the falsity of the theory can be inferred from empirical evidence, and this inference is a purely deductive one." is indeed a paradox. While, as we discussed, the facts by themselves are not sufficient for proving a theory correct, it is a necessary part of the verification of a theory. As well, we should not always rely on facts for disproving a theory for it is possible to design the proper experiments that would support a theory. Unless we have a clear logical structure of a theory we cannot prove nor reject a theory on the basis of observation. "Why is it reasonable to prefer non-falsified statements to falsified ones?" asks Popper. "The only correct answer is the straightforward one: because we search for truth (even though we can never be sure we have found it), and because the falsified theories are known or believed to be false, while the non-falsified theories may still be true. The principle of empiricism can be fully preserved, since the fate of a theory, its acceptance or rejection, is decided by observation and experiment – by the results of tests." (K.R. Popper).

Although an experiment is a necessary part of the validation of a theory, it is only one part of it. There is no clear description of the relation between scientific theory and observation from Popper's point of view.

Pierre Duhem in "Physical Theory and Experiment" discusses the role of experiment in the validation of a physical theory. "An experiment in Physics is not simply the observation of a phenomenon; it is, besides, the theoretical interpretation of this phenomenon. An experiment in physics is the precise observation of phenomena accompanied by *interpretation* of these phenomena; ... An experiment in physics can never condemn an isolated hypothesis but only a whole theoretical group." (P. Duhem). Duhem distinguishes two types of experiments. The experiment for testing a hypothesis is accompanied, in his opinion, by interpretations. In my opinion such experiment is designed to fit an already existing theoretical interpretation of a hypothesis. The other type of experiment Duhem calls 'an experiment of application'. "This experiment does not aim at discovering whether accepted theories are accurate or not; it merely intends to draw on these theories. [There] is nothing to shock logic in this procedure." (P. Duhem). He also thinks, as does Popper, that the most convincing way to criticize a theory is to present experiments that instill and justify doubts in the theory. "A physicist disputes a certain law; he calls into doubt a certain theoretical point. How will he justify these doubts? How will he demonstrate the inaccuracy of the law? From the proposition under indictment he will derive the prediction of the experimental fact; he will bring into existence the conditions under which this fact should be produced; if the predicted fact is not produced, the proposition which served as the basis of the prediction will be irremediably condemned." (P. Duhem). Although experiment is a powerful argumentation in debate, for we are used to believing our eyes, just as well as facts by themselves do not prove a theory, facts by themselves do not disprove either.

For testing the theory of elasticity an experiment has been designed in order to find a limit of elastic relations that can be deduced from the standard equation of elastic deformation for a bar in tension. From the equation of elastic deformation, $e = NL/EA$, a partial derivative is obtained, $\mathrm{d}e/\mathrm{d}A = -NL/EA^2$, in order to find the area of cross-section which corresponds to rapid changes of deformation and failure of elastic behavior. A tensile test was conducted on a number of specimens of the same length but different areas of cross-section. The experiment did not produce the expected result. It did not show a correlation between A and the rate of change of deformation. A change was then made to the description not because facts pointed to a solution, but rather because the author believed in the logic of a description that should produce the necessary result. The new equation of deformation $e = N/ER$ and its derivative equation $\mathrm{d}e/\mathrm{d}R = -N/ER^2$ allowed us to calculate the theoretical limits with the equation of elastic stability. Now the test did support the theory. The role of such a test is supplemental. In the course of testing it appears that in some cases the actual limit demonstrated in the test corresponds to the calculated value while in others the limit appears to be less than that predicted by calculation. The conclusion was that the latter limit corresponds to the limit of the material. In this instance when an experiment was made for testing a logically correct theory the experiment was useful not only for validating the theory, but for correcting the physical concept as well. The explanation–conclusion that I made is that the actual limit of elasticity of a structure is relative. It can be the limit that depends on the geometrical stiffness of the structure and is calculated with the equation of elastic stability, or it can be the limit of elasticity of the material that is obtained experimentally. The real limit is the lesser value. I did not doubt my theory because some facts did not corroborate with the logical structure. Rather I looked for the possible explanation of the facts in the frame of a new theory. If no possible explanation would come to me then maybe the useful theory would have to be abandoned. Negative experiments do not necessarily disclaim a theory.

Duhem refers to the physical science as to a whole system like an organism that is too complex to obtain proof, but only may give some degree of validation. "A 'crucial experiment' is impossible in physics. Between two contradictory theorems of geometry there is no room for a third judgment; if one is false, the other is necessarily true. Do two hypotheses in physics ever constitute such a strict dilemma? Shall we ever dare to assert that no other hypothesis is imaginable? Unlike the reduction to absurdity employed by geometers, experimental contradiction does not have the power to transform a physical hypothesis into an indisputable truth . . . " (P. Duhem).

My point of view is that not a crucial experiment confirms one theory while excluding others, but rather the logical system connecting proposition and inference in the domain of possibility of physical existence may confirm a theory. The construction of NLTE shows that it is possible to validate a theory within a new logical system and within accepted postulates of physics in such a definitive exclusive way that no other theory would stand by unless a different system or different facts are introduced in the theory. The selection of the physical entities that place the function of deformation in the domain of the tangent function allows us make an *inductive* judgment on the values of the unobservable theoretical rate of change of deformation. "Are certain postulates of physical theory incapable of being refuted by experiment? Certain fundamental hypotheses of physical theory cannot be contradicted by experiment, because they constitute in reality definitions, and because certain expressions in the physicist's usage take their meaning only through them." thus P. Duhem answers his own question.

NLTE uses not only the values of the tangent function, but also its domain in the interval $0 < \tan\alpha < 4$ for demarcating the domain of the law of elasticity. Furthermore we can maintain that this interval of the tangent function is the domain that demarcates the existence of possible physical relations, for it is illogical and against our physical experience to assume that a physical stable relation may exist while the tangent function representing the relation shows the tendency to infinite increase. Thus the declaration is that only this certain interval of the tangent function is logically justified as the domain of existence of a physical relation. It also has validation in observation and experimental facts. Thus the deformation increases slowly to some extent. After that the decrease of cross-section follows with a rapid increase of deformation and elastic failure or destruction. For some functions the rate of change corresponds to the pattern of the tangent function. For some other propositional functions the derivative function is not a tangent function. The domain of those systems nevertheless is the same interval of the tangent function. In our example in this work with constant force the selection of the tangent function as the derivative function that coincides with the domain for the system is justified. The values of inference are strictly determined by the tangent function. The function that satisfies this requirement has mathematically strict demand for selection of its variable-concepts. In order to be physically justified the function should have experimental support. For the phenomenon of elasticity the basic physical entities we deal with are: a whole body with its resistance to changes, that is an observable entity; the external forces, which are observable entities; and the change of the geometry of a body, which is an observable and measurable entity. However we cannot construct a function based on these observables. It is wrong to think that there is a functional connection between the external force that causes deformation, and the response of the body to the force. The response of the body depends on its geometry. The external force can be distributed or concentrated, but the body responds to it as a geometrical whole. The body has its own laws of distribution of deformations and internal elastic forces. A connection exists between distribution of deformations and the distribution of elastic forces. The deformations of the whole body and its parts are observable and measurable. However, the internal elastic force or stress is, in principle, unobservable. We know of the existence of such force based on the general postulate of physics that if work is present then should be a force that perform this work. The magnitude of this unobservable entity can be found theoretically.

Does this mean that that it is impossible to have a crucial experiment that allows exclusively one from a number of theories? The answer is that such a procedure of elimination is possible, but it is rather theoretical than experimental elimination. NLTE does not contradict the universal law of physics for energy conservation, and it is within a strict logical–geometrical structure that secures its results. On the other hand, we can get rid of the unobservable entity (elastic force) by substituting it by the corresponding observable entities. Thus, we have the equivalent description of force that can be measured, $F = Ee_aR_a$, where e_a is the actual deformation of the body and R_a is the actual geometrical stiffness of the body. For obtaining optimal geometrical stiffness of the body we measure the deformation from the actual observable force in the actual body and thus reach the ideal condition for given external force and the experimental finding of deformation, $R_o = \sqrt{e_aR_a/C_s}$. Optimal geometrical stiffness is projected from the experimental data of actual observables and a deductive conclusion on the behavior of similar structures according to the description given by the tangent function. It is possible to move from the description with unobservable internal forces to one with observable geometrical changes instead.

Concerning the universal principles or hypotheses accepted as universal principles in physics, Pierre Duhem wrote "We are certain that we shall never be led to abandon it because of a new experiment, no matter how precise it is." A theory that is based on a consistent logical structure cannot be abandoned because we have not designed a proper experiment to support such physical theory. The conclusion Pierre Duhem made is "Certain physical theories rest on hypotheses, which do not by themselves have any physical meaning. Good sense is the judge of hypotheses which ought to be abandoned." This is not the conclusion to which NLTE led us. Besides good judgment, whatever it means, certain logical inductive and deductive conclusions should be validated with experiments designed for this purpose.

Sharp criticism of justificationism and dogmatic falsificationism appears in the paper "Falsification and the methodology of scientific research programmes" by Imre Lacatos.

> According to "justificationists" scientific knowledge consisted of proven propositions. Having recognized that strictly logical deductions enable us only to infer (transmit truth) but not to prove (establish truth), they disagreed about the nature of those propositions (axioms) whose truth can be proved by extra logical means. *Classical intellectualists* (or "rationalists" in the narrow sense of the term) admitted very varied – and powerful – sorts of extra-logical "proofs" by revelation, intellectual intuition, experience. These, with the help of logic, enabled them to prove every sort of scientific proposition. *Classical empiricists* accepted as axioms only a relatively small set of "factual propositions" which expressed the "hard facts". Their truth-value was established by experience and they constituted the *empirical basis* of science. In order to prove scientific theories from nothing else but the narrow empirical basis, they needed logic much more powerful than the deductive logic of the classical intellectuals: *"inductive logic"*. It turned out that both theories are equally unprovable. *(I. Lacatos)*

As an objective criterion of impossibility to reach the definitive proof of scientific theory Lacatos refers to the existence of a number of credible geometrical systems and different views on mechanics of relations in the physical universe. We discussed already that the existence of non-Euclidean geometry does not eliminate Euclidean geometry. If our argument fits a world of Euclidean plane geometry then it is exactly what we can use. If the argument requires the accuracy and simplicity of Newtonian physics, then again, it is what we will use. We work within a certain system of geometrical axioms and physical laws. The rules of logical and physical determinism are the same for any developed geometrical and physical systems.

The definition of deductive logic and the deductive inference are well understood, though the method is applied mainly in one direction, i.e. if a proposition is correct then the inference is also correct. In my work I found it more rewarding to start with a necessarily true inference that corresponds to the tangent function, and then construct the corresponding proposition. This system is logically a deterministic system. But how can we assign a physical truth to this system? Experimental verification of the constructed proposition is necessary for validating the proposition and conclusion but it is insufficient for proving that our system is a consistent part of some general physical law. In order to prove that we should demonstrate the connection of the unobservable entities in the propositional function to the basic physical laws and definitions. For example, in the function of elastic deformation the elastic force is an unobservable entity. In order to prove a function that has such unobservable entity we have to prove its existence through covering it with a universal law in our physical system of knowledge. The existence of the elastic force is proved by its correspondence to the law of conservation of energy and definition of work. In this quite reliable and definitive construction of the physical law of elastic relations we use both principles of deductive logic and principles of inductive logic. We use the principles of deductive logic in combining the two-level system of functions. One of the functions, i.e. the function describing the rate of change of deformation, is a function that can be referred to the Euclidean logical–geometrical system. Inference from the corresponding equation of deformation is certain because it belongs to Euclidean plane geometry. The proposition is certain because it is a part of this deductive logical system. As well as the deductive structure is validated with the known Euclidean logical system with its basis in plane geometry, so is the inductive proof based on the connection of our system of elastic relations to the yet untouchable physical foundation of conservation of energy and the definition of work. As concerns the observable entities – they do not need proof, deductive or inductive, for they can be measured. These terms rather need to be properly selected in order to be fit into the constructed logical system.

Let us return to the criticism of justificationism and falsificationism in the Methodology of Scientific Research Programme. According to falsificationism a theory builds its propositions by deriving them from other propositions. Propositions cannot be derived from facts. As well, statements of a theory cannot be proved with facts. However they can be disproved with facts. Science progresses by overthrowing theories in presenting facts falsifying them. There are different trends in falsificationism, in the rules of demarcation of a theory and rules for elimination of a theory. The theory can be saved only with help of auxiliary hypotheses, which satisfy certain well-defined conditions. The final judgment of a satisfactory theory belongs to an empirical criterion that endured test with time.

The methodology of scientific research programmes consists of methodological rules: Some rules are designed to show common pitfalls of the theories (negative heuristic), while other rules show possible ways for obtaining positive results (positive heuristic).

> Our considerations show that the positive heuristic forges ahead with almost complete disregard of "refutations": It may seem that it is the "verifications" rather than the refutations which provide the contact points with reality. *(J.A. Kourany)*

In "Testing Theoretical Hypotheses" Ronald N. Giere wrote: "I shall be arguing for a theory of science in which the driving rational force of the scientific process is located in the testing of highly specific theoretical models against empirical data." Giere views theoretical models as free from logical contradictions. They also have no empirical content and tell us nothing about the world. However, when the elements of a model are identified with the elements of real systems we may claim that the real system exhibits the structure of the model. Giere calls this system a theoretical hypothesis. This hypothesis is either true or false. "The simplest form of a theoretical hypothesis is the claim that a particular, identifiable real system fits a given model. Contrary to what some philosophers have claimed, one can have a science that studies but a real single system." (R.N. Giere). And further he gives his definition of scientific theory.

> What then is a "theory"? [Most scientists] think that "theories" have empirical content. This is a good reason to use the term "theory" to refer to a more or less generalized theoretical hypothesis asserting that one or more specified kinds of systems fit a given type of model." *R.N. Giere*

How is one to test a theory? Giere says: "In the language of testing, no test of a theoretical hypothesis can be completely reliable. No model of science that places the relations between evidence and hypotheses within the probability calculus will prove to be true." Why then does he propose the models? "It is similarly with prediction. Hypotheses that are useful in making the reliable predictions are desirable, but not because this makes them any more likely to be true. Rather, there is pragmatic value in being able to foresee the future adequate." It is difficult to convince others of the value of your methodology if the author condemns it part by part.

7.10 On the logic of truth-function

Mathematics is often identified with deductive reasoning. In "The Philosophy of Mathematics" Stephan Korner wrote "A true or false compound proposition, the components of which are also either true or false, is a truth-functional proposition (briefly, a truth-function) if and only if, the truth or falsehood of the compound proposition depends only on (is a function of) the truth or falsehood of the components."

The conclusion in our research of the compound physical function is contrary to Stephan Korner' position. Different functions may have true components and from the mathematical point of view these functions are correct, but the test of these propositional functions includes not only tests of the components of the function but also a test of the correctness of the logical structure of which the proposition is a part. It was shown that the function of deformation might have all true components of geometry. A function may have the right concept and magnitude of elastic force. It may have the correct magnitude of the resulting function describing deformation. At the same instance this so-called truth-function can give a false description of the relations between the components. For testing the relations we need to construct a logical system of functions. According to Korner: "Since a class of truth-functions, which are identically true, is well defined and since it can be decided by routine methods whether any given truth-function is or is not identically true, there is no need to construct a deductive system the postulate and theorems of which would embrace all identically true truth-functions and no other propositions. ... Within the domain of truth-functional logic the difference between obviously and not-obviously analytic (logically necessary) propositions is no longer important. All truth-functional tautologies are eligible premises and some are needed as such." (S. Korner).

The position in this work concerning the exclusiveness of a truth-function is very different. The propositional compound functions cannot be identical from the physical and logical points of view. In order to separate truth-function from falsehood-function it is necessary to construct an inductive–deductive system within the domain of possible physical existence that would test the propositional function. And it is not enough to test the components of a function to pronounce its truth-value.

7.11 On the logic of classes

> The set consisting of all sets is itself a set, and thus an element of the set of all sets itself. These sets may strike us as being somehow abnormal as compared with normal sets of classes, which do not contain themselves as elements. *(B. Russell)*

In "Principia Mathematica" Russell suggests "Whatever involves all of a collection must not be one of the collection." Here again, the point of view in my work is quite different. The set of strength characteristics of a structure contains both the characteristics of the material parts of the structure and the characteristics of the structure as a geometrical whole. Comparison of the strength characteristics of the members of such an 'abnormal set' leads to the conclusion on the final adequate characteristic of strength of a structure. If the logic is general for mathematics as well as for physics, then it should be noticed that the principles of logic in mathematics are not sufficient for physics.

7.11.1 On the logicist systems

One may now analyze the following citations concerning the relations between mathematics and logic from the work of Bertrand Russell.

- "Any logicist system has to be judged both from the mathematical and from the philosophical point of view."
- "Yet the mere mathematical perfection of the system is not sufficient to validate a logicist philosophy of mathematics."
- "Every logicist system draws its list of postulates and the list of inference from the logic of truth-function, the extended logic of classes, and the logic of quantification."
- "The fields of pure mathematics and of rational theology share the characteristic of being a priori. Yet the two can be clearly distinguished and neither of them is reducible to the other. It is similarly possible that pure mathematics shares it's a priori character with logic, and that they yet are irreducible the one to the other."

- "The search, however, for such general characteristic [of analyticity] as would cover both logic and mathematics has, so far, been unsuccessful, which is perhaps the less surprising, in view of the pragmatic principles included among the postulates of logicist systems, especially principles which are almost indistinguishable from empirical hypotheses about the universe."
- "Mathematics and logic are two separate a priori sciences."
- "That geometrical statements are not unique has been demonstrated by the construction of self-consistent non-Euclidean geometries."
- "What can be confirmed or falsified by perceptions – experiments and observations – is not a geometry or any set of a priori statements but a physical theory using the geometry."
- "If mathematics is to be restricted – entirely and without qualification – to the description of concrete objects of a certain kind, and logical relations between such descriptions, then no inconsistencies can arise within it: precise descriptions of concrete objects are always mutually compatible."
- "To show that a system of propositions – e.g. – the theorems of a mathematical theory – is internally consistent is to show that it does not contain two propositions one of which is the negation of the other or a proposition from which any other proposition would follow."
- "It is obviously possible to exhibit a selector-function for a class consisting of a finite number of finite classes. It is impossible to pick out one element from an infinite number of finite classes or from an infinite number of infinite classes because selection is a feature of perceivable or constructable objects or processes."
- "It has been demonstrated by Gödel that every embodiment of classical mathematics in a formalism must be incomplete; there is always mathematical truths, which are not embodied in theorem-formula."
- "Traditionally, the task of logic has been conceived as that of providing criteria of correctness for inferences by making explicit the rules which are conformed to by correct inferences and violated by incorrect ones."
- "The apparent failure to derive mathematics from logic led thinkers – as Kant had done – on such intuitions as were backed up by the particular subject-matter of mathematical construction."

8. Conclusion

In this work I have tried to accomplish two main goals. One was to devise a reliable method for the optimization of structures. The other was to build a logical structure of the theory that allows mathematical and experimental verification of the physical theory.

While the problem of optimization is connected with the construction of a reliable theory, the method of building a theory has its own independent importance. My first task was accomplished in 1986 after more than a dozen years of consideration and working on the problem. In 1997 I received a U.S. Patent on "Method of and Apparatus for Optimization of Structures". Although the method has its roots in the new theory, it appeared before the logical structure of the theory in its present form had been developed. The theory is described for the first time in this book after the definition of the theory as Non-Linear Theory of Elasticity came up.

On my way to these goals many side problems appeared that needed to be solved, and they have been considered. Some other problems remain untouched and require revision from the new scientific position. The purpose of this book is not to solve all engineering problems. Such a universal task is not available for the individual. It is continuous in time and space. It is not a textbook either, though it provides comprehensive instructions for practicing engineers. The purpose of the book is, rather, to present a new scientific foundation for structural analysis and design and a new method for the optimal structural design process. Below is a short concluding journey through the theory where the ideas are the milestones.

For selecting the optimal reliable dimensions of a structure, which is a common engineering problem, we have to compare the elastic behavior of similar structures differing only in dimensions. For that we need a description of elastic relations that we can apply to a set of similar structures. The question is, can we use the current linear equation of deformation or the equation for the elastic line for this purpose?

The best way to examine the equation of deformation from this point of view is to present the equation as a curve. Then the difficulties with the current equation become apparent immediately. If we select the same scale for the absolute values of the function, i.e. the deformation, and for the independent variable, i.e., in the current equation, the cross-sectional characteristic, then it is impossible to draw the graph, because the value of deformation can be hundreds or a thousand times less than the area of cross-section. In such a graph not only the difference between the values of deformation disappears but the deformation itself disappears as well. The phenomenon of elasticity becomes invisible. This is not, as one may think, a formal mathematical problem but a problem that concerns the adequate presentation of the physical phenomenon.

If, on the other hand, we choose equal scales for the significant figures, both for the independent variable and for the function, then this gives us the possibility to see changes of deformation depending on the changes of the cross-sectional characteristic. However, this leads to another problem. The curve does not correspond to the equation of deformation in the linear theory. The rate of change of deformation calculated with the equation and the tangent to the curve have different values. As the graph describes the elastic behavior of a set of similar structures more satisfactorily with regard to the physical observation than the equation of deformation, the next task was then to change the equation. We can select a different independent variable, one that relates to the deformation and that can be at the same dimensional level as the deformation.

In what other way can we describe the phenomenon of elasticity? To that end, all characteristics representing the phenomenon should relate to deformation. The fixed body that is subjected to the action of the external force experiences changes. The deformation of the body depends not only on the force but on the resistance of the body to deformation as well. The elastic force that originates in the body is proportional to the deformation. Thus, the main characteristics that represent the phenomenon are the elastic force, the resistance to deformation, and the deformation, $D = F/ER$. The resistance to deformation ER depends on geometry and on the elastic properties of the material. The part of the resistance that depends on geometry was defined as the geometrical stiffness R. The geometrical stiffness of a body first of all depends on the ratio of the cross-sectional characteristic of a body to its length. The geometrical stiffness also includes the effect of all specific features of design that are represented with the experimental coefficient K, $R = KA/L$.

This description of deformation is an adequate description of a phenomenon of elasticity. The graphical presentation of the equation of deformation corresponds in this case to the mathematical description. The graph helps visualizing the relations in a set of similar structures with different geometrical stiffnesses.

The curve/function exists in the domain of the tangent function. This domain has two intervals, i.e. the interval of rapid changes and the interval of slow changes of deformation. The limit of elasticity for a structure is characterized by a rapid increase in deformation. The rate of change of deformation can be described with the derivative function. The rate in the vicinity of $\tan 76^{\circ} = 4$ can be used for determining the structure-specific critical stress that is generated by the geometry of the structure. The desirable geometrical stiffness can be calculated using the derivative equation, which is defined as the equation of elastic stability, $F/ER^2 = C_s$, where $C_s = \tan \alpha$ is the coefficient of elastic stability selected by the designer.

The description of geometrical stiffness for a particular structure gives the designer the possibility to change geometrical parameters for achieving the optimal geometrical stiffness, $R_o = \sqrt{F/EC_s}$.

How can one prove that the physical theory is not an artificial mathematical construction, but true to the facts, and that we can rely on the results obtained with the equations?

The non-linear theory of elasticity is designed in a way that allows proof. The equation of deformation in the non-linear theory has a mathematical inference. This inference is a function that is known to be correct, i.e. it is the tangent function in its two intervals. Each point of the basic function has a derivative that is known to be true. This makes the function known theoretically in its full domain. The basic equation was tested experimentally numerous times as well. The rate of change of deformation may have experimental support in the interval of rapid changes. The linear theory has no such logical structure. It cannot be proved. However the linear theory was disproved. The principle of physical determinism allows only one correct description with the exclusion of the other descriptions.

For proving a logical system of connected physical functions we need both experimental and logical verification. For establishing the domain of a phenomenon we need theoretical and experimental verification.

Theoretically the domain of possible existence of stable physical relations is determined within the interval of the tangent $0 < \tan\alpha < 4$. No matter what character a physical function has and no matter what its derivative is, the rate of change of the function should not exceed $\tan\alpha = 4$, as beyond that the magnitude of the tangent rapidly increases towards infinity. Such increase designates physical instability.

The coefficient of elastic stability is the universal criterion for testing the physical existence of stable relations.

Empirical verification of this particular limit shows that the limiting characteristic of a structure can be of different physical origins. It can be the limit of the material or it can be the limit generated by the geometry of the structure, whichever is lower. Knowledge of the limit of elasticity for a particular structure and knowledge of the elastic behavior for a structure depending on the position of that structure in a family of similar structures allows us to select the optimal geometrical stiffness of a structure with a high degree of mathematical certainty.

We are living in a world of structures and machines. The safety and cost of the structures, the use of natural resources and energy, the use of transportation and machinery, and consequently the environment, depend on the knowledge of minimal reliable dimensions of machines and structures. The non-linear theory of elasticity allows us to improve technology.

9. Recapture of the central ideas

(1) The non-linear theory of elasticity describes the elastic relations in a structure that is considered as geometrical whole.

(2) The equation of deformation in NLTE refers both to the individual structure and to a set of similar structures.

(3) NLTE is justified with the methods of definitive logic, i.e. inductive and deductive, and use of auxiliary methods of mathematics, experiments, and measurements.

(4) Definitive logic establishes the necessity of a two-level logical structure that includes a basic equation and its derivative. Without such logical structure it would be impossible to reach definitive results and justify or falsify a theory.

(5) The system of functions connected by mathematical inference is necessarily a deductive procedure. This system should be considered in the physical and mathematical domain of the function.

(6) The domain of possible physical relations is obtained logically with reference to the approximate interval of tangent function, $0 < \tan \alpha < 4$. If the rate of change of a physical function exceeds the limiting value it is incompatible with physical existence, for beyond this limit the magnitude of the tangent increases rapidly towards infinity.

(7) The conclusion regarding the domain of physical function is made with the inductive method of definitive logic. The inductive method refers justification of a specific law to the knowledge of some universal law or definition accepted in science. In the case of establishing the domain of physical existence it is the knowledge of the character of the tangent function that provides certainty of conclusion.

(8) The auxiliary methods in physical theory do not provide justification, but have their own functions.

Mathematics is an instrument in science that allows manipulation of variables, calculations, and transformation of a function from one level of relations to another. By itself mathematics does not prove or falsify a physical theory.

(9) Experimental methods are designed in order to obtain data for supporting a theory. Before an outline of a theory is made there cannot be a meaningful experiment.

(10) Some of the terms/variables in a physical function are observable, others are unobservable theoretical terms. Observable properties can be obtained with empirical methods. Unobservable properties, on the other hand, cannot be observed but can be calculated. Their certainty and values depend on the logical connection of the unobservable terms to some observable properties. Thus, the elastic force is an unobservable entity. Its value can be obtained,

for instance, by using Hooke's law of equivalence of stress and strain.

(11) Hooke's law belongs to the type of laws that is used for establishing the physical equivalence of observable (ε) and non-observable (σ) terms describing the same physical condition, such as the characteristics of the elastic field, $\sigma = E\varepsilon$. The unobservable term in a theory can be traced to the cause of the phenomenon. The inductive inference can be made to connect the cause, that is, the external force, and the effect, that is, the description of deformation.

Bibliography

Acherkan, N.C., Spravochnik Mettallista, Vol. 2, ed. 1960, Mashgiz, Moscow.

Ackermann, R.J., Modern Deductive Logic, ed. 1970, Anchor Books, Doubleday & Company, Inc.

AISC, Manual of Steel Construction. Vol. 1, Load and Resistance Factor Design, ed. 1994.

Avallone, E.A., and T. Baumeister III, Handbook for Mechanical Engineers, ed. 1987, 1996, McGraw-Hill.

Ayer, A.J., Language, Truth and Logic, ed. 1952, Dover Publications, Inc.

Birger, I.A., and Y.G. Panovko, Prochnost, Ustoichivost, Kolebania, Spravochnik Vol. 1, ed. 1968, Mashinostroenie, Moscow.

Campbell, H.E., and P.F. Dierker, Calculus, ed. 1975, Prindle, Weber & Schmidt.

Carnap, R., An Introduction to the Philosophy of Science, edited by M. Gardner, ed. 1995, Dover Publications, Inc.

Castell, A., A College Logic, ed. 1935, Macmillan & Company.

Chen, W.F., The Civil Engineering Handbook, ed. 1995.

Cohen, M.R., Preface to Logic, ed. 1963, Meridian Book, The World Publishing Co.

D'Abro, A., The Rise of the New Physics, Vol. 1, ed. 1951, Dover Publications, Inc.

Dewey, J., Logic. The Theory of Inquiry, ed. 1938, Henry Holt & Company, Inc.

Einstein, A., Ideas and Opinions, ed. 1985, Crown Publishers, Inc.

Eshbach, O.W., and M. Souders, Handbook of Engineering Fundamentals, 3rd ed., 1975, John Wiley & Sons.

Faupel, J.H., and F.E. Fisher, Engineering Design, 2nd ed., 1975, Wiley-Interscience.

Gödel, K., On Formally Undecidable Proposition, 1931, reproduced in "From Frege to Gödel", J. van Heijenoort, ed. 1967.

Gödel, K., Russell's Mathematical Logic, Kleene Stephen

Gödel, K., What is Cantor's Continuum Problem?, Kleene Stephen

Holton, G., Thematic Origins of Scientific Thought, 1973, Harvard University Press, Inc.

Hume, D., An Enquiry Concerning Human Understanding, reproduced in "Classics of Philosophy", L.P. Pojman, ed. 1998, Oxford University Press, Inc.

Kant, I., Prolegomena to any Future Metaphysics, reproduced in "Classics of Philosophy", L.P. Pojman, ed. 1998, Oxford University Press, Inc.

Kourany, J.A., Scientific Knowledge. Basic Issues in the Philosophy of Science, ed. 1987, Wadsworth.

Laudan, L., Progress and Its Problems, 1977, University of California Press, Ltd.

Leibniz, G.W., Discourse on Metaphysics, reproduced in "Classics of Philosophy", L.P. Pojman, ed. 1998, Oxford University Press, Inc.

Love, A.E.H., A Treatise of the Mathematical Theory of Elasticity, ed. 1892.

Morley, A., Strength of Materials, 11th ed., 1954, p. 1, Longmans Green & Co.

Popper, K.R., The Logic of Scientific Discovery, ed. 1968, Harper & Row.

Ratner, L., Method of and Apparatus for Optimization of Structures, U.S. Patent, 1996.

Ratner, L., Method of Optimization of Structures, Experimental Techniques **23**, 1999, SEM, Inc.

Ratner, L., Deflections Indicate Design Similarity, Machine Design, 1986, Penton/IPC.

Russell, B., Introduction to Mathematical Philosophy, 1918.

Sandifer, C.E., Mathematics, from Mark's Standard Handbook for Mechanical Engineer, McGraw-Hill.

Spinoza, B., Selections edited by J. Wild, ed. 1958, Charles Scribner's Sons.

Suppe, C.F., editor, The Structure of Scientific Theories, ed. 1977, University of Illinois Press.

Tamboll, A.R., Steel Design Handbook, LFRD, ed. 1997.

Thewlis, J., Encyclopedic Dictionary of Physics, ed. 1962.

Thompson, J.M.T., and G.W. Hunt, A General Theory of Elastic Stability, ed. 1971.

Todhunter, I. and K. Pearson, A History of the Theory of Elasticity and of the Strength of Materials, ed. 1884.

Whitehead, A.N., Science and the Modern World, ed. 1969, Macmillan & Company, Inc.

Williams, L.P., The Origins of Field Theory, ed. 1980, University Press of America, Inc.

Subject Index